All About
Bioinformatics

All About Bioinformatics
From Beginner to Expert

Yasha Hasija

Department of Biotechnology,
Delhi Technological University,
Delhi, India

ELSEVIER

ACADEMIC PRESS
An imprint of Elsevier

Academic Press is an imprint of Elsevier
125 London Wall, London EC2Y 5AS, United Kingdom
525 B Street, Suite 1650, San Diego, CA 92101, United States
50 Hampshire Street, 5th Floor, Cambridge, MA 02139, United States
The Boulevard, Langford Lane, Kidlington, Oxford OX5 1GB, United Kingdom

Notices
Knowledge and best practice in this field are constantly changing. As new research and experience broaden our understanding, changes in research methods, professional practices, or medical treatment may become necessary.

Practitioners and researchers must always rely on their own experience and knowledge in evaluating and using any information, methods, compounds, or experiments described herein. In using such information or methods they should be mindful of their own safety and the safety of others, including parties for whom they have a professional responsibility.

To the fullest extent of the law, neither the Publisher nor the authors, contributors, or editors, assume any liability for any injury and/or damage to persons or property as a matter of products liability, negligence or otherwise, or from any use or operation of any methods, products, instructions, or ideas contained in the material herein.

ISBN: 978-0-443-15250-4

For information on all Academic Press publications visit our website at https://www.elsevier.com/books-and-journals

Publisher: Stacy Masucci
Acquisitions Editor: Linda Versteeg-Buschman
Editorial Project Manager: Michaela Realiza
Production Project Manager: Sajana Devasi P K
Cover Designer: Vicky Pearson Esser

Typeset by TNQ Technologies

Working together
to grow libraries in
developing countries

www.elsevier.com • www.bookaid.org

Contents

What is bioinformatics?

1

1.1 Introduction

Bioinformatics is an interdisciplinary life science field. This is a field which deals with the collection and efficient analysis of biological data. In other words, it is a recently developed science which uses information to understand biological phenomenon. The research in bioinformatics is regarded as a domain which encompasses expanding, complex, and large datasets. It is a part of computational biology which addresses the necessity of managing and interpreting the data, massively generated in the past decade by genomic research. Bioinformatics is the discipline that integrate the biotechnology and information technology, interpretation and analysis of data, genomics convergence, development of algorithm and modeling of biological phenomena. Bioinformatics is a wide encompassing branch and is therefore difficult to define. For some it is still an ambiguous term encompassing biological modeling, system biology, biophysics and molecular evolution. Whereas for others it is simply computational science applied to a biological system.

Bioinformatics entails the usage of high technology solutions in biological experiments through a variety of computer programs. Bioinformatics is becoming a vital part of biology. Bioinformatics uses image and signal processing technology to derive essential knowledge from vast volumes of data. In genetics, genome sequencing and mutation analysis may be significantly important. Information plays a vital role in the research for biologists and the creation of gene ontologies. It plays an essential function in the study of gene and protein influence. Bioinformatics tools help in explaining and contrasting molecular biology evidences. It helps in recognizing the biological processes and biological networks that are active in the biological system. We need this in structural biology to simulate and forecast molecular activity.

Bioinformatics is a proliferating area which is currently in the foreground of science and technology. Various institutes all over the world are heavily investing (especially because of the pandemic) in possessing, transferring and exploiting the data for future development. It is a valuable ticket at present, and bioinformatics learners would thrive from the demand for jobs in private sector, in government and academia.

There are several elements of science that contribute to bioinformatics. It also implies to biological molecules and thus includes knowledge of the fields of

All About Bioinformatics. https://doi.org/10.1016/B978-0-443-15250-4.00012-5

molecular engineering, molecular biology, statistical mechanics, biochemistry, thermodynamics, molecular evolution, and biophysics. The use of computer science, mathematical, and statistical principles are needed in the field. Bioinformatics is at the intersection between experimental and theoretical research. Its not just about "mining" data or modeling, it is about analyzing the molecular environment that drives life from the perspective of evolution and mechanisms. It is genuinely cross-disciplinary and is evolving. Like genomics and biotechnology, bioinformatics is evolving from applied to fundamental research, from creating tools to creating hypotheses.

Bioinformatics, Computational biology, and bio information infrastructure are sometimes used interchangeably.

1. **Bioinformatics** relates to the methods, observations, and data storage utilized in the genomic era.
2. **Computational biology** requires the usage of software to analyze biological processes better.
3. **Bio information infrastructure** includes all the information software, computational methods, and networking networks supporting biology. The latter dissertation may be seen as an informational scaffold for the first two.

1.2 History

Bioinformatics was first properly established 50 years back. Although the term "Bioinformatics" was coined by Ben Hesper and Paulien Hogeweg in 1970 (Hesper and Hogeweg, 1970), but the tracks of its emergence go back to 1960s with the efforts put by Margaret Oakley Dayhoff, Russell F. Doolittle and Walter M. Fitch (Chang et al., 1965). The contribution of Margaret is so important to this field that former director of NCBI David J. Lipman called her "the mother and father of bioinformatics". There was need to compare and analyze a huge amount of protein sequences or amino acids sequences from different organisms computationally as it was manually impractical to handle such large data. This is what lead Margret O. Dayhoff, "the first Bioinformatician" and her colleagues at the National Biomedical Research Foundation in compiling the first ever "Protein Information Resource" (PIR), stating that "analysis of protein was the starting point for bioinformatics" (Dayhoff et al., 1974). They successfully organized the protein sequence data into various groups and subgroups according to the requirement. Further contribution to the development of bioinformatics was given by Elvin A. Kabat in the 1970s by his extended analysis of protein sequences of comprehensive volumes of antibody sequences. Further in 1974, George Bell and his associates initiated the DNA sequences collection into the GenBank, with the objective of contributing to the theoretical background to immunology. The primary version of GenBank was being prepared in 1982−1992 by Walter Goad's group (Burks et al., 1987). Consequently, The DNA Data Bank of Japan (DDBJ) and European Molecular Biology Laboratory

(EMBL), "the world's first nucleotide sequence database" were also made in 1984 and 1980 respectively.

The first conceptualization of Bioinformatics in Switzerland was in the early 1980s. Swiss bioinformaticians developed software to compare genetic nucleotide sequences, created programs for the study of experimental peptide and protein results, invented computer tools for three-dimensional modeling structures of proteins, and created databases of protein details. These individuals participated in the field of bioinformatics and contributed to biology science in general. In 1998, Swiss bioinformatics became unified. The current five Swiss bioinformatics groups combined to create the SIB (Swiss Institute of Bioinformatics), a charitable organization.

However, the most important development in these databases was the incorporation of the web-based searching algorithms which helped researchers in their queries. GENEINFO was the resulting computer software developed by David Benson, Lipman and associates. The software was made available through NCBI (National Center of Biotechnology Information) web-based interface. Also, NCBI was made available online in 1994 along with the tool BLAST (Altschul et al., 1990). Afterward, several major databases which are still in usage like PubMed (1997) and Human Genome (1999) came into existence.

Bioinformatics tools are growing evermore prolific and are increasingly expected to replicate all results. To help students understand evolution more accurately, professors are integrating this theory into biology students' curriculum. Synthetic biology, systems biology, and whole-cell modeling have emerged due to the ever-increasing complementarity between computer science and biology (Hagen, 2000).

1.3 Biological databases

A biological database is a complex, extensive and complete structured collection of biological data arranged in computer readable form which enhances the search speed and retrieval. Biological databases have appeared as a response to the massive amount of data provided by the low-cost DNA sequencing technologies. The first database to develop was GenBank, which is a compilation of all the accessible DNA and cDNA sequences.

Previously databases were perceived somewhat different. However, over the course of time the term "biological database" has become a default concept. The data is directly submitted to the biological databases for organization, indexing and optimizing the data. The databases help students, scientist and researchers to find, discover and analyze the related biological data by making it accessible in a format which can be read and used on software's. This is the primary purpose of these databases, storing, managing and retrieval of biological information. A range of information can be retrieved from these biological databases like binding sites, molecular actions, biological sequences, metabolic interactions, motifs, protein families, molecular action, homologous and functional relationships etc. A lot of Bioinformatics work is based on data collection and manipulation. Any of these provide both "public" and "private" database references. Having those databases available to different computers allows it far simpler for more users to communicate these databases efficiently.

Biological databases may be broadly classified into PRIMARY, SECONDARY and DERIVED databases. And then there are further distribution. Primary databases contain only sequential and structural information. These can also be called as archival databases. They are loaded with experimentally generated data such as protein and nucleotide sequences. Experimental data are directly submitted to database by scientist or the researcher. After the database accession number is given, the primary database data will never be changed: they become the part of the scientific record. Few examples of primary biological databases are: GenBank and DDBJ for genome sequences, EMBL, Swiss- Prot and PIR for protein sequences and Protein Databank for protein structures.

Secondary databases are those which constitutes data/information from primary databases or the analyzed result of the primary databases. The primary databases often have minimal sequence annotation information. A much more post processing of the sequence data is required to convert raw data of sequences into more sophisticated biological information. This implies the necessity for databases containing computationally analyzed sequencing data which is obtained from primary databases. Hence secondary database comes into the picture containing information of the results of the analysis of primary data. Secondary databases are highly curated and consists of more valuable information in comparison to the primary databases. The databases comprise data such as signature sequence, active site residues and conserved sequences. Few examples of secondary databases are: UniProt KB, Motif databases, PDB and InterPro etc. (Fig. 1.1).

Composite or derived databases are the amalgam of the primary and secondary databases. To enter the data as input in the database, it is first compared and then sorted on the basis of the desired parameters. Primary databases are source for extracting initial data and is then combined in conjunction based on specific parameters. They consist non-redundant data. Examples of composite databases are: OMIM (Online Mendalian Inheritance in Men) and Swissport. Also, there are databases which are specialized for a particular research interest, for e.g. HIV sequence database, Flybase and Ribosomal Database project.

1.4 Algorithms in computational biology

Computational biology/Bioinformatics are interdisciplinary areas which are concerned with employing computers capacities to address biological interests' problems. The two terms are used interchangeably, but there's a consensus formed between the two. Bioinformatics refers or focuses on the activities which gives attention on developing and utilizing computational tools for the analyses of the biological data. Whereas Computational biology refers to those activities which mainly works on constructing or developing algorithm leading to address the biologically relevant problems.

Biological Data

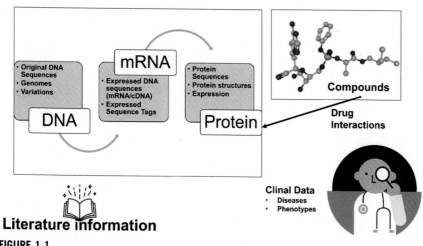

FIGURE 1.1

Biological information to data.

Up until recently biologists have not had access to such massive quantity and quality of data generated and stored in different databases discussed above. Over the past 2 decades, unprecedented technical advancement has been made in producing biological data, techniques like microscopy, next generation sequencing, high throughput techniques etc have contributed to data explosions. Researchers are producing datasets that are so enormous that it has become impossible to analyze, manage and make proper use of the data to understand biological process and their relationships. This is what led to the introduction of the various algorithms in the field. An algorithm is a process or description about how to solve a problem. Modern day computer's ability to perform and store billions of calculation and processes makes it possible to use the amount of data generated not only in just biology but any other field. The computational biology algorithms have several uses including prove or disapprove a certain hypothesis.

The process of creating algorithms that resolves biological significant issues, in computational biology comprises of two steps. First phase is to raise an interesting biological question and to build a model of biological reality that makes it possible to articulate the question as a computational problem. Secondly, construct an algo which will be able to solve the formulated computational problem. The primary step needs biological reality knowledge, while the latter requires algorithmic theory knowledge. The algorithm quality is a combination of its space assumption and running time and the answers of biological relevance it produced. Data scientist frequently torture the various data structure in order to reduce the ambiguity of space and time. This approach has explicitly benefited researchers well in the field. Hence having a working knowledge of basic computational algorithm is of paramount importance to bioinformaticians or researchers in the field, also their expertise in the development of novel algorithm would have a strategic advantage in both academia and industry (Fig. 1.2).

There are many algorithms already existing in the field, helping in the current research. Some of them are Dynamic programming: Needleman Wunsch (Global alignment) and Smith waterman (Local alignment), Hidden Markov Models, Principal Component Analysis Clustering, Phylogenetic tree construction, machine learning applications (SVM, neural network), microarray data analysis, protein secondary structure prediction and many more.

Global and Local sequence alignment uses our understanding of a organism's proteins to understand more about the proteins of other organisms. Next HMM are used for sequence modeling or model a DNA sequence. In HMM, the probability of happening of an event is dependent on its previous state. This model uses a probabilistic finite state machine in which a letter is emitted depending upon the present state probability and then move to next state. The next state can be possibly e equal to the original one. In gene regulation networks, they formed because of the different protein's interaction in an organism. The various proteins regulate each other and, depending on the structure of their interactions, the cell type is determined (Crombach and Hogeweg, 2008).

1.5 Genetic variation and bioinformatics

Genetic variations are the modifications (changes) in the chromosome sequences. Variation is also the reason why two individuals of the same species having similar characteristics, but are not identical. It is the engine of evolution which enables

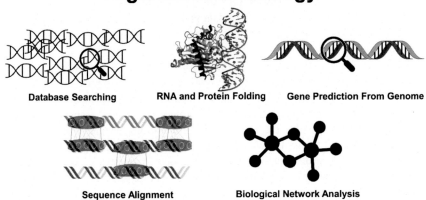

Algorithms in Biology

Database Searching   RNA and Protein Folding   Gene Prediction From Genome

Sequence Alignment   Biological Network Analysis

FIGURE 1.2

Challenges in biology solved through algorithms.

organisms to conquer the environmental challenges they meet (Stoletzki, 2008). It may be both damaging and effective in the development of efficient mechanisms in cell factories to tackle changes and survive. In the evolution of biotechnology, efforts have been made to make use of genetic variation to our advantage in order to produce strains with beneficial phenotype. It is also stated as the variation in the sequences of DNA among people within a population and it happens in somatic cells as well as germ cell (egg and sperm cells). The only difference that exists is the variation in germ cell can be inherited through one person to another person, hence impacting population dynamics and subsequently evolution. The main cause of variance are recombination and mutations. Mutations are said to be the original source of variance causing permanent alteration in DNA sequence. It can be harmful, beneficial or neutral to the organism. The other main reason for genetic variation is recombination. Every organism has a combination of genetic information from their parents. Thus, recombination happens when these genetic materials combine or say homologous DNA strands are crossed and aligned. SNPs (single nucleotide polymorphism) are the genetic variation which is very common among people. Every SNP reflects a variation in the DNA base A, G, C, T of a person's genome. They occur on average once in every 300 bases and are also present between genes. The core priority area of modern medical research is studying the effect of SNPs on human health. In more than 1% of population Single nucleotide substitution can be observed. Numerous algorithms have been applied to evaluate the impact of SNPs mainly focused on the human genotype data analysis which classifies variations either diseases-causing or neutral, tolerant or intolerant and deleterious or neutral. Which implies that the genetic variation would either expected to have no impact or inflict some significant negative effects on the phenotype. The one downside to these algorithms is that they are classifiers build on existing knowledge and it is well said that biology is the science of exceptions, presently scientific community has able to uncover only the tip of the ice burg representing biological phenomena. Therefore, these tools are built on the assumptions which we have onboard presently. These tools are used for predicting disorders and are mainly used for diagnostic purposes. There are many available tools and databases for predicting the effects of SNPs. Some of them are Variant Effect Predictor (VEP) which evaluates the impact of variants on genes, protein sequences, transcript etc., SIFT (Sorting Intolerant from Tolerant) is sequencing homology-based tool which filters tolerant amino acids and also determines whether the substitution of amino acid in protein would have a phenotypic effect. DbSNP is the SNP database of NCBI, which contain information on non-polymorphic, microsatellite and deletion/insertion forms.

Bioinformatics in Genetic variation covers the following areas;

(a) Latest algorithms and software development for genetic variance analysis and application with pipelines and visualization tools.
(b) Genetic variations analysis in the genome; DNA and single nucleotide polymorphisms (SNPs); techniques to assess numbers of people with a disease; study large-scale data sets.

(c) It involves studies of data sets, and study of recent methodological advances in the area of genetics.

(d) Involve in genetic variance identification, functional annotation, pathway simulation, and analytical methods built for different sequencing platforms.

1.6 Structural bioinformatics

"Structural bioinformatics is a subset of Bioinformatics that deals with the prediction and analysis of 3D structure of Macromolecules such as DNA, RNA, and Proteins." And the second thing which comes is why understanding the structure of macromolecule is important. The reasons are first: structure determines function, so learning structure helps in understanding of function. Secondly the Structure is more conserved than the sequence, hence enabling identification of a much more distant evolutionary relationship. Thirdly understanding the structural determinants enables the design and modification of proteins for industrial and medical benefit. The structural bioinformatics field and concepts related to it offers not only a way of coordinating views about sequence-structure-function questions but also a mechanism for detecting unobserved behavior and proposing novel experiments (Konings et al., 1987; Schuster et al., 1994).

Proteins are essential components of cells of living species. The structural specificity of a protein is related to the role of the protein. Protein structure visualization is a subject of recent biochemistry research and is an essential method for structural bioinformatics. Most often used is: **Cartoon**: This illustrates the secondary structure variations for the protein. Besides, as α-helix is often interpreted as a form of a screw, β-strands is also described as arrows, and loops as arcs. **Lines:** Each atom is depicted by thin lines and allows for a lower cost of data in a visualization. **Surface:** In this visualization, one can see how the molecule appears. **Sticks:** There are covalent connections between amino acid atoms in proteins. This kind of cluster graph strategy is most widely used for visualizing relationships between amino acids (Fig. 1.3).

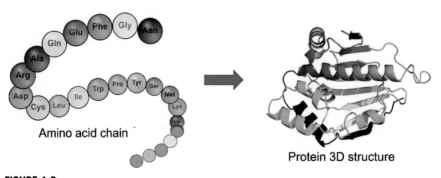

FIGURE 1.3

Amino acid chain to protein 3D structure.

A substantial majority of bioinformatics study focuses on the estimation, interpretation and simulation of protein 3D structures. Proteins first 3D structure (that of myoglobin) was experimentally determined in 1958, through X-ray diffraction. However, in 1951, Pauling and Corey set the first milestone in the protein structure prediction. As in other fields of biological sciences, it is now possible to predict secondary and tertiary structure using computer calculations and that too with varying degrees of certainty. High-throughput methods have given the knowledge required to relate protein structures to their results. This structural and therapeutic details can be valuable for bioinformatics applications in medical science. Computerized visualization of the protein models provides insights into biological processes that cannot be appropriately described otherwise.

Though advances in 3D structure prediction field are vital, it is significant to know that proteins are a dynamic network of atoms rather than being static. With many advancements in biophysics, force fields have been designed to explain atom interactions between themselves, which enabled the development of tools to model protein molecular dynamics in 1990s. Even though tools were developed and theoretical methods were available, but because of the huge computational resources needed, it remained rather complicated to execute molecular dynamics simulation. It can be explained by the example that it required weeks for the calculation of a microsecond simulation of a protein, using a supercomputer with 256 CPUs. Despite several improvements in the modern computers power like the use of GPU (Graphics processing units) or graphics card, it is still not accessible to perform molecular dynamics simulations on reasonable time scale. But yes, increasing computational power in conjunction with the increasing data have made the process a little bit convenient.

1.7 **High-throughput technology**

High-throughput sequencing methods have become important in the field of genomic and epigenomic studies. With the advent of increasingly advanced sequencing tools, the amount of DNA Sequencing Approaches has risen tremendously. It has revolutionized the molecular biology field by enabling high-scale whole genome sequencing and also a wide variety of experiments to study the internal cell workings explicitly at the RNA or DNA level. The data generated is the findings of widespread molecular project like gene expression analysis, multiple projects of genome sequencing, protein-protein interactions and analysis of genomics. They are compiled and deposited in a number of databases.

High throughput sequencing is generally divided into two classes: RNA seq and Genome sequencing. In the latter one, sequencing of fragmented genomic DNA is done and reads sequence is used for the assembling of the while genome. But on the other side, RNA-seq attempts to read the sequence taken from the RNAs. In both the cases, reads may be paired end or single end. For RNA seq, reads are produced from both the ends of the longer fragmented RNA or DNA. While choosing

High throughput technology the user should consider the quality control issues, sample collection, along with the biological hypothesis being tested.

It is a technology developed as an alternative to microarray. Although high throughput technology is also comparatively more costly than the microarray, it still has many benefits for the evaluation of the factors that influences the gene expression regulation. For e.g.: microarray is limited to the model organism to which the microarray has already been built while HTS may be extended to a non-model organism. But High throughput methods are now getting cheaper and are likely to replace even fingerprinting methods including analysis of traditional clone library. High throughput sequencing methods provides the capacity for the detection of rare phylotypes particularly, effectively offering quite reliable estimates of relative abundance and assessment of diversity indices. The key benefit of HTS is that it produces good quality gene expression data sets. There is a need of specialized tool for viewing, storing, indexing, organizing and analyzing biological and computerized data. Thus, bioinformatics is the bridge between computational and biological sciences, which can provide a deeper insight into this field. Data sets of HTS are both complex and high dimensional in nature. It is quite challenging both computationally and algorithmically to integrate such data with various other data sets in order to attain profile of a diseases completely. To integrate HTS data, network -based approaches have the ability to incorporate data from various sources while ensuring results that are relevant. There are many multidisciplinary programs like molecular tumor boards that put together biologist, physicians and bioinformatics, which helps in addressing the challenges of translating the data that are important to health care providers and patients. Substantial computational power is needed by the related algorithmic approaches. High Performance computing (HPC) offers resources which can be exploited by computer/bioinformatics researchers. Some of the resources offered by HPC are cloud computing platforms, GPU (Graphic Processing Unit), or clusters. Every resource is different in terms of performance, technology, ease and scalability of implementation, and cost. The table shows example of High Throughput applications which uses GPU, cloud or other resources (Table 1.1).

Table 1.1 High throughput computing devices.

S.NO.	HPC	High throughput applications
1.	GPU	Read mapping, Process-intensive task (RNA-seq alignment)
2.	Clusters	Self-Organizing Maps (SOM) Exome analysis workflow
3.	Cloud	Used by private and public clouds like 1000 genome project and international cancer genome consortium.

1.8 **Drug informatics**

Drug Informatics relates to the combination of computer techniques and pharmacy expertise to discover and examine drugs. Drug-Informatics is the study of relationships between drugs, their mechanisms, and structures, focusing on medication awareness and improving the quality of life. It is not manually possible for healthcare professional to hold all of the information which is required to provide medical care with safety and efficacy aided by the scientific knowledge present today. The scenario can further worsen as there is huge increment in the complexity and volume of data regarding mechanisms of diseases generated through genomic revolution. The solution to this whole mess is the acquiring of thorough technologies and techniques for the management of the data. And that's where drug informatics comes into picture as it enfolds the area where these technologies and techniques affect the use of drug data in a commercial, clinical or research setting.

It is the sum of all data and information generated throughout the life cycle of a drug. The delivery of this information from lab to patient is not direct. It involves drug informatics to provide specific information related to the patient. The sources of drug informatics have been classified into 3 categories: primary, secondary and tertiary. Primary level of literature includes data from the research lab of a university or a pharmaceutical company or even the clinical observations from a hospital. It also includes unpublished and published data from various journals but not every published literature in a journal is said to be primary literature, such as editorials or review articles are not said to be considered as primary literature. Secondary level of literature consists of the data created as a repercussion of the use of data originally accumulated for non-clinical purposes like pharmacy benefit reimbursement and management. It also consists of sources that either abstract or index primary literature, with the intention of leading the individual to the appropriate primary literature. Tertiary literature contains material which has been compiled and condensed by the editor or author to provide a short and simple overview of the subject. Few examples of this consist of journal review papers, compendium textbooks and general data which can be downloaded on the internet (Fig. 1.4).

During the evolution of the drug informatics and its practices the healthcare sector was faced significant challenges: rising prices, excessive loss rates, and disgruntled patients and providers. Then entered the information technology applications in the field during the 1960s which focused on clerical and financial systems. With the advent of effective network technology and personal computers in 1980s, there came the introduction of more clinically oriented computing systems for healthcare. It is still a field which is still new particularly in comparison with other disciplines of medical. Drug informatics is an exponentially growing field with the application of information technology and computer science to health and medical data. It encourages the technology use as an essential resource to efficiently manage, analyze, organize data on the use of drugs on patients. The fundamental aim of the field is to disseminate two categories of information knowledge-based information: consists healthcare scientific literature and patient specific information: generated

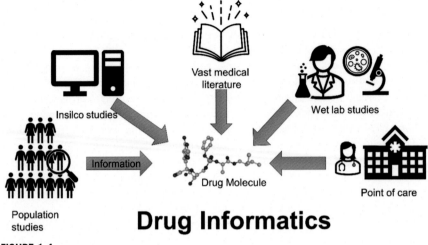

Insilco studies

Vast medical
literature

Wet lab studies

Information

Drug Molecule

Point of care

Population
studies

Drug Informatics

FIGURE 1.4

Schema of drug informatics.

during taking care of the patient. Policy makers and other leaders in health informatics are now inclined toward the improvement of the safety and quality of health care systems while at the same time lowering the cost.

1.9 System and network biology

Network and Systems Biology is the analysis of complex interconnecting systems through the comprehensive method. Network and systems biology attempts to explain the interaction network between hundreds of various biological molecules concurrently. Biological science is undergoing two increasingly evolving phenomenon: first is the increased computer capabilities and advancement in software tools, secondly advent of High throughput biological data on proteomes, transcriptomes, genomes and so on. The biological field has become a data intensive field due to which computer science and biology are now parallel to each other along with fields like algebra, chemistry, physics and statistics. The integration of these fields has contributed in the development of network biology, bigdata and other divisions of biology. The aim of the network biology is to comprehend cells or organisms as a whole at different stages of mechanism and functions. The difficulties of analyzing and evaluating massive biological networks and molecular data is now faced by system biology. The types of data in system biology are sequence data: DNA sequences, Molecular structure data of DNA and RNA extracted by using NMR and X-ray crystallography, Gene expression data, binding sites and domains data, protein -protein interaction and metabolic pathways data. Generation of network is done after receiving data from different data types. After this multivariate analysis is done of the network, using regression analysis, PCA (Principal Component Analysis) and Clustering. Some of the applications of network algorithm in system biology are: Function prediction: system level approach for determining the functions of an entity which is not known is carried out by network creation of that entity along with the unidentified and identified entities, Protein Complex detection: done using Y2H (yeast two hybrid system) and Affinity purification- Mass Spectrometry, Analyzing evolution, Drug development, Disease diagnosis and Interaction prediction (Crombach and Hogeweg, 2008; van Hoek and Hogeweg, 2006; Lindenmayer, 1968).

1.10 Machine learning in bioinformatics

Machine learning is a technique for analyzing large amounts of data and developing predictive tools. While logical and computational algorithms pass the input data through some logical rules or mathematical operations and provide the desired output, on the other hand machine learning algorithms are trained on data consisting of input and their outputs and formulate the rules for calculating the outputs during training. Thus, machine learning can be used to calculate a predictive output from new input data. Machine learning techniques are based on probabilities and statical functions, and they attempt to fit the training data into a mathematical model or function. Machine learning has become critical for solving complex problems in the field of genomics. Machine learning has emerged as a savior as the rapid evolution of high throughput technologies outpaces the development of techniques capable of making sense of all available data. As with any other field of science and technology, life science has embraced an Artificial Intelligence-based approach to dealing with the avalanche of data generated (Minsky and Papert, 1969) (Fig. 1.5).

In general, machine learning algorithms can be classified into three categories: supervised learning, unsupervised learning, and reinforcement learning. When both the input and output variables are known, supervised learning is used; this technique is primarily used to develop predictive tools based on known data. Unsupervised learning is used in cases where the output results are unknown, but we attempt to cluster data points based on their similarities or dissimilarities. Unsupervised learning is used to discover correlations and variations in data or to identify outliers. Reinforcement learning is used in situations where a dynamic environment requires multiple outputs or a series of multiple outputs for specific inputs. Algorithms are trained to select the best output based on a reward and punishment system, where the correct output is rewarded and the incorrect output is punished. At the moment, machine learning is being aggressively applied to nearly every problem in bioinformatics. As a result, machine learning has become a necessary tool in the toolbox of every bioinformatician. Chapter 9 of this book contains a more detailed discussion of machine learning algorithms and their applications.

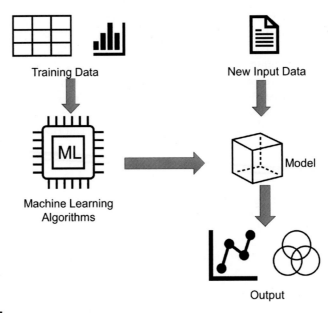

FIGURE 1.5

Schema of a machine learning problem.

1.11 Bioinformatics workflow management systems

A bioinformatics workflow management system is built primarily to compile and implement a sequence of bioinformatics steps. As new pipelines are being developed to analyze the data generated from various sources, standardization of these pipelines becomes essential for the reproducibility of results. Pipelines are first divided into their components then each components are built individually such that their inputs and outputs are in standard form and can be integrated independently when require.

There are numerous workflow frameworks. Experts from various fields like have developed some of the leading research workflow models. Such workflow is based on an abstract description of how a computational pipeline continues in the context of a directed graph. Each framework usually features a visual interface, enabling users to construct complex programs by joining the input and outputs of various components form various pipelines without becoming fluent in computer programming.

Various workflow management systems exist like;

KNIME (Konstanz Information Miner): It is a free, open-source data processing framework. Many tools combine to build complex structures for machine learning and data mining under the modular term "Lego of Analytics."

Online HPS: OnlineHPC is an online toolset that offers high-performance computing and job flow services to broad agencies. An online group of scholars working on computer systems science has developed.

Galaxy: Galaxy tools help scientists without computer programming to handle computational biology results.

UGENE: UGENE is a software application for bioinformatics. It operates on Windows, Mac OS X, or Linux operating systems. UGENE software helps geneticists examine genomic details such as DNA genomes, genome descriptions, phylogenetic trees, and assembly files. The data can be processed both locally and online (e.g. a lab database).

Gene Pattern: GenePattern is a publicly accessible computational biology opensource software kit produced and built at the Large Institute, built to replicate new genomic analyses by researchers.

1.12 Application of bioinformatics

Bioinformatics is a multidisciplinary area which develops software's tools and methods for analyzing the biological data. The advances in the both computer sciences and molecular biology over the past 30–40 years has contributed in the development of bioinformatics. Bioinformatics key areas include sequence alignment, genomics, proteomics, molecular phylogenetic and many more. It is intended to set up an automated archive that catalogs genomes and protein sequences from single-celled species to multicellular ones. This approach enables three-dimensional structures of complicated molecules to be studied and interpreted. Bioinformatics is a mixture of mathematical, bio-chemical, and numerical approaches to analyze human, biochemical and biophysical evidence. Bioinformatics is concerned with gene sequences and their encoded details.

The bioinformatics field has become important for basic molecular biology and genomic research along with having a significant influence on various fields of biomedical science and biotechnology. Bioinformatics is being used in almost every field of biology starting with medicine the various applications of it are:

Drug Designing: Changes in drug designing happened within a decade when the first 3-D protein structure was illustrated. The information of target protein 3-D structure was incorporated into the drug design procedure. In the process of drug designing the structure of protein can influence the whole process at any stage. Various online tools and software's have been made to make for better accessibility and results. For e.g.: Auto dock.

Personalized medicine: With the advancement of the pharmacogenomics area, clinical medicine can become more personalized. This is an analysis about how genetic inheritance of an individual influences the body reaction to drug. Till now, physicians were using trail and error methods to the determine the right medication to treat a single patient, since people with the same health symptoms can show a wide variety of reactions to the same therapy. In last decade personalized medicine has emerged which allows doctors to prescribe best dosage and available drug therapy from the start by analyzing the genetic profile/biomarkers of the patient (Anderson et al., 2008).

Gene Therapy: It is the approach used in the curing, preventing and giving treatment of the diseases by changing the expression in the genes of a person. With the use of bioinformatics tools, the possibility for the use of genes itself to cure diseases have become the reality now.

Microbial Genome Applications: The emergence of full genome sequences and their ability to provide deeper understanding of the microbial ecosystem as well as the capabilities, could have far reaching and wide consequences for energy, industrial, environmental, and health applications. After the genetic material analysis of these organisms, researchers may start to known these microbes in their primitive form and identify genes that provides them their remarkable ability to live under harsh conditions.

Evolutionary studies: Genome sequencing of all the three organisms, archaea, bacteria, and eukaryote signify that evolutionary studies could be carried out in the pursuit of determining the tree of life and bioinformatics can surely help using phylogenetics studies.

Computational approaches for bioinformatics expand knowledge and strategies for cellular and molecular levels into information for moving up to the level of inanimate structures such as habitats. Genome sequences offer a way of recognizing the usual biological mechanisms, clarifying the malfunctioning of genes contributing to diagnosing diseases and designing new medicines. There are so many applications of bioinformatics in various fields of agriculture, microbial genome and medicine etc. Using bioinformatics will help biologists and researchers to excel in the experiments and will also helps in extending their skills in data processing more reliably and quickly.

References

Altschul, S.F., Gish, W., Miller, W., Myers, E.W., Lipman, D.J., 1990. Basic local alignment search tool. J. Mol. Biol. 215, 403−410. https://doi.org/10.1016/S0022-2836(05)80360-2.

Anderson, A., Chaplain, M., Rejniak, K., Fozard, J., 2008. Single-Cell-Based Models in Biology and Medicine.

Burks, C., Fickett, J., Goad, W., 1987. GenBank status report. Science 235, 267−268. https://doi.org/10.1126/SCIENCE.235.4786.267-C, 1979.

Chang, M.A., Dayhoff, M.O., Eck, R.v., Sochard, M.R., 1965. Atlas of Protein Sequence and Structure. National Biomedical Research Foundation.

Crombach, A., Hogeweg, P., 2008. Evolution of evolvability in gene regulatory networks. PLoS Comput. Biol. 4.

Dayhoff, M.O., Barker, W.C., McLaughlin, P.J., 1974. Inferences from protein and nucleic acid sequences: early molecular evolution, divergence of kingdoms and rates of change. Orig. Life 5, 311−330.

Hagen, J.B., 2000. The origins of bioinformatics. Nat. Rev. Genet. 1 (3), 231−236. https://doi.org/10.1038/35042090.

Hesper, B., Hogeweg, P., 1970. Bioinformatica: een werkconcept. Kameleon 1 (6), 28−29. Dutch, Leiden: Leidse Biologen Club.

Konings, D., Hogeweg, P., Hesper, B., 1987. Evolution of the primary and secondary structures of the E1a mRNAs of the adenovirus. Mol. Biol. Evol. 4.

Lindenmayer, A., 1968. Mathematical models for cellular interactions in development II. Simple and branching filaments with two-sided inputs. J. Theor. Biol. 18.

Minsky, M., Papert, S., 1969. Perceptrons.

Schuster, P., Fontana, W., Stadler, P., Hofacker, I., 1994. From sequences to shapes and back: a case study in RNA secondary structures. Proc. Biol. Sci. 255.

Stoletzki, N., 2008. Conflicting selection pressures on synonymous codon use in yeast suggest selection on mRNA secondary structures. BMC Evol. Biol. 8.

van Hoek, M., Hogeweg, P., 2006. In silico evolved lac operons exhibit bistability for artificial inducers, but not for lactose. Biophys. J. 91.

Introduction to biological databases

2.1 Introduction

A database is a large collection of well-organized data designed to remain intact over time. Databases are typically associated with a piece of computer software that allows users to alter (i.e., update), search for, and retrieve specific sets of data that are kept within a computer-based framework. The DBMS software, which stands for "database management system," collaborates with the user application, and the database to store, manage, and change data in such a way that it may be utilized to get information that is of use. To obtain data, and create tables or objects, insertion, deletion, and modification of data in a database, are done through queries written in query language specific to a DBMS. Some of the examples of DBMS include MySQL, Oracle, and PostgreSQL.

Among different sorts of databases, the ones comprising the databases applicable to natural sciences like molecular science and bioinformatics are called biological databases. In the current situation, the significance of biological databases can be perceived from the accompanying focuses:

- Huge volumes of raw data, such as raw sequencing, proteomes, and other types of data, are being generated at a very rapid pace due to many technological advancements in molecular research and proteomics and the cheap cost of high-throughput genome sequencing. As a result, the capability and handling of this incredible data are among the most pressing issues facing genomics at the moment.
- It contributes to the development of customized medicine by helping prescribe the most effective drug.
- This approach allows for the modification of either the patient's or the person's DNA to treat genetic diseases.
- As of now, biological databases have become the focal point of bioinformatics. Through the different information mining devices, all-natural data can be effortlessly gotten to, consequently saving time, assets, and endeavors.
- Additional uses include the manufacture of bioweapons, the investigation of evolutionary processes, the improvement of agricultural yields, and the increase of the nutritional value of food.

All About Bioinformatics. https://doi.org/10.1016/B978-0-443-15250-4.00001-0

- With the emergence of the machine learning era, NLP is being utilized for automated curation of data, and thus biological databases allow knowledge discovery (when the raw data is entered, there is hidden information that is not known beforehand; entry into the database helps uncover those undiscovered connections between a piece of information entered, which is called knowledge discovery). This facilitates the uncovering of new biological insights from raw data.
- Multiple-point access is possible, as is convenient retrieval of publicly accessible data.
- improved indexing (for global access).
- reduced redundancy as data is curated computationally and manually.

2.1.1 Characteristics of biological data

The majority of other forms of data are substantially simpler in comparison to the complexity of biological data. The definitions of this kind of data have to be able to show both the intricate substructures of the data and the relationships that exist between them to guarantee that no information is lost during the process of modeling biological data. This is the only way to prevent any information from being overlooked. Not just in a hierarchical, binary, or tabular layout, but the data model should be able to display any amount of complexity included in any data schema, relationship, or schema substructure. For instance, the National Center for Biotechnology Information (NCBI) biological data model interprets a biological sequence as a straightforward integer coordinate system, to which many types of data may be connected. The coordinate system is connected to a vast amount of information, including the sequence in which amino acids are found in the body.

The quantity and range of data are quite different from one another. Therefore, there must be some degree of adaptability in the handling of data kinds and values. A single piece of data may need to be represented by more than one data type. This is since biological data structures often include exceptions. In addition, the sorts of data that are gathered for various species and genome projects sometimes overlap with one another.

In biological databases, schemas are subject to constant modification. The majority of relational and object database systems do not now support the addition of new fields to the schema. Therefore, nucleotide sequence databases like GenBank will re-release the whole database with new schemas rather than making incremental improvements to the system as required, which may seem invisible to the user.

People who work with biological data often need to review prior versions of the same data. As an example, GenBank makes use of Accession. The version number in the flat file entries helps keep tabs on the protein sequences. In addition to this, the version contains a GI number, which denotes a certain sequence. While the accession number does not change throughout updates, it does get a new GI number if there is a modification to the protein sequence. GenBank is a public database that focuses on nucleotides; however, when nucleotides are translated, proteins are created from the nucleotides (Gligorijević and Pržulj, 2015).

2.2 Types of databases

Most databases can be put into one of three main groups: primary databases, secondary databases, and composite databases.

2.2.1 Primary database

Primary databases, which are also known as archival databases, are made up mostly of datasets that were generated through experiments. These datasets include things like nucleotide and protein sequences as well as information on how macromolecules are assembled. Users can augment this fundamental information by adding functional annotations, references, and linkages to other databases. The material was entered into the primary database by the researchers themselves. As soon as the information is received, it is included in the scientific record and assigned a unique number that is referred to as an "accession number." Types of primary databases include the following:

1. **Primary nucleotide sequence databases**: The three most significant databases that save raw nucleic acid sequences and make them accessible to users are GenBank (Bilofsky and Christian, 1988), EMBL (Kanz et al., 2005), and DDBJ (Okubo et al., 2006). Users may access these sequences via these databases. Users may also access the sequences by navigating through these databases. You may access GenBank in the United States by going to the internet gateway for the National Center for Biotechnology Information. GenBank is a database that stores genetic sequences. The headquarters of GenBank may be found in the United States of America. In their respective regions of the globe, the European Molecular Biology Laboratory (EMBL) and the DNA Databank of Japan (DDJB) may be located. The European Molecular Biology Laboratory, which is also known as EMBL, is in the UK, while the DDJB is in Japan.

2. **Microarray/Functional genomics databases:** The study of trials that make use of high-throughput technologies to assess transcripts, proteins, and metabolites is what is known as the field of functional genomics. These investigations need a great deal more information regarding the design, sample, and procedures utilized than genomics does since they are highly dependent on the circumstances. For functional genomics, we want databases that adhere to both the current and emerging data quality requirements. There are a few functional genomic databases that concentrate on microbial genomes, and there are a few different methods that may be used to build up functional genomic databases. However, functional genomic databases are not yet particularly prevalent.

3. **Protein sequences and structure databases**: PIR-PSD and SWISS-PROT (Boeckmann et al., 2003) are both examples of databases that hold the protein sequences that have been determined. PIR-PSD is located at the National Biomedical Research Foundation (NBRF) in the United States of America, while SWISS-PROT is housed in Switzerland at the Swiss Biotechnology Institute (SBI). During the development of the PIR-PSD, a collaboration between the PIR, the MIPS (Munich Information Center for Protein Sequences, Germany), and the JIPID (Japan International Protein Information Database, Japan) was essential. The PIR-PSD is now a complete object-relational database management system that is well annotated and has no unnecessary components (DBMS). The PIR-PSD is unique in that it organizes protein sequences into groups by using the concept of "superfamilies" as its organizing principle. In addition, the PIR-PSD sequence is partitioned into groups according to the sequence motifs and homology domains that it contains. Homology domains might be compared to the building blocks of evolution, while sequence motifs refer to functional locations or regions that have not undergone any changes. The approach of categorization makes it simpler to see how the sequence, function, and structure are all interconnected with one another.

2.2.2 Secondary database

The original database's information is transferred to a secondary database. A secondary sequence database contains data such as the conserved sequence, signature sequence, and active site residues of protein families discovered by repeated sequence alignment of a set of related proteins. The PDB entries are stored in a secondary structure database. There are listings for all alpha proteins, all beta proteins, and so on, according to how they are created. These also include information about a protein's recurring secondary structural motifs. Secondary databases created by various academics and housed at their labs include SCOP, created at Cambridge University; CATH, created at University College London; PROSITE, created by the Swiss Institute of Bioinformatics; and eMOTIF, created at Stanford (Rother et al., 2005).

1. **Protein families, domains, and structure databases:** Protein databases are now indispensable to the practise of modern biology. The structures, functions, and, most importantly, the sequencing of proteins are the subjects of a significant amount of current research. When investigating a novel protein, the investigation often starts with a search via several databases. By comparing individual proteins or whole protein families, we may learn about the connections that exist between the proteins in a genome or between the proteins of different species. This is far more information than we might get by examining a single protein on its own. InterPro, PROSITE, SCOP, CATH, and the NCBI Conserved Domain Database are all included here (CDD).

2. **Protein sequences and functional information databases:** The functional information consists of a collection of data that the functional head needs to successfully perform and administer the function. Because this knowledge is solely helpful for performing that one function, it cannot be put to use in any other context. This information will be used by a manager to plan and handle the work at hand.

3. **Nucleotide (Genes/Genomes) sequence and annotation databases:** The process of obtaining information on the structure and function of a protein or gene from a raw data set is referred to as "genome annotation." This is accomplished by the use of a variety of mining methods, including analysis, comparison, estimate, and precision. It is essential to annotate the genome since sequencing the genome or DNA results in the creation of sequence information that provides no insight into how the system functions. After the genome has been sequenced, the data must next be annotated to provide further details about the genome's structure and the way it functions, includes NCBI, Ensembl, etc.

2.2.3 Composite database

The necessity to search in many places is eliminated when using a composite database since it integrates the information from many different major databases. The search methodology used by each composite database is based on a separate primary database as well as a unique set of criteria. In the composite database, there are many different ways to search for the information you need. Researchers are granted unrestricted access to the nucleotide and protein databases that are stored on the massive, high-availability, redundant array of computer servers that are maintained by the National Center for Biotechnology Information (NCBI), which is the organization that is responsible for hosting these databases. In addition to this, a link is provided to the Online Mendelian Inheritance in Man Database, which has information on the proteins that may be associated with inherited diseases. There are situations in which a data collection may function as either a main or secondary database. For instance, primary peptide sequences may be easily uploaded to the Uniprot database. In addition, Uniprot can collect protein clusters from primary peptide sequences. It may also have the automated and manual explanations from TrEMBL and SwissProt (Bateman et al., 2017).

2.3 Models of databases

The logical framework that is used for the storage of computational results via DBMS is referred to as the database model. The kind of database model that is used is a major factor in determining how effectively data can be retrieved and stored. Earlier database models consisted primarily of two-dimensional data tables or a single file containing fields and their values. But because data has grown so much in the last few decades, database models have become more complicated and linked.

A few of the database models we'll cover in the next sections are listed below.

- Flat File
- Hierarchic
- Network
- Entity-Relationship
- Relational

At least one presentation is possible within the context of every information base administrative structure. The optimal structure is determined by the consistent connection of the application's data as well as the application's requirements, which include exchange rate (speed), unswerving quality, practicability, adaptability, and cost. The perfect structure is determined by these factors. Even though products can provide support for more than one model, the vast majority of information database management systems are designed to revolve around a certain information model. Any given coherent model may be actualized using a variety of different physical information models. Because the choices that are made have such a large impact on how the program is run, the vast majority of database programming will provide the user with some level of control over the process of tuning the physical execution (Fig. 2.1).

A model is not only a means of organizing information; rather, it also outlines a variety of actions that may be carried out on the data. These activities include: For example, the social model characterizes tasks such as "select (venture) and join." Even though the results of these actions might not be clear in a certain question language, they are still the building blocks on which a question language is made.

2.3.1 Flat file

The most fundamental kind of database is one in which information is kept in the form of a file or a two-dimensional array (table) and is organized into columns, rows, and values. In this architecture, the columns are responsible for defining the various fields, while the rows hold the data of a single record and are linked together by a shared ID. The relational model evolved from its predecessor, the flat file model. The fields and values that are part of GenBank entries can be thought of as an example of a basic flat file paradigm in biological databases.

2.3.2 Hierarchical model

The information included in this database model is structured in a manner that is reminiscent of a tree, and each of the other nodes is linked to the one that serves as the tree's root. The hierarchy begins with the root data and develops in the shape of a tree when more child nodes are added to the nodes that are already there and make up the parent nodes. Every child node is connected to exactly one parent

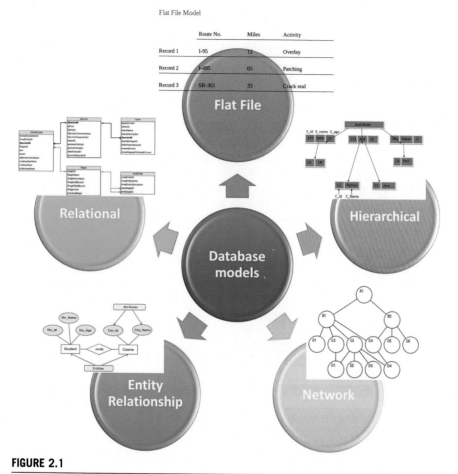

FIGURE 2.1

Some common database models in DBMS.

node. The data in a hierarchical model are laid out in a structure that resembles a tree, and there are relationships between the various categories of data.

Example of a hierarchical database in the biological sciences: the cell type is the root of the biological function network database; each cell has multiple organelles and cytoskeletal elements; each component is involved in multiple pathways and functions; and there can be more levels of hierarchy (Fig. 2.2).

2.3.3 **Network model**

A network-based extension of the hierarchical model is shown here. The data in this model is structured more like a graph, and a node can have more than one parent node. The data in this database model becomes more interconnected as more

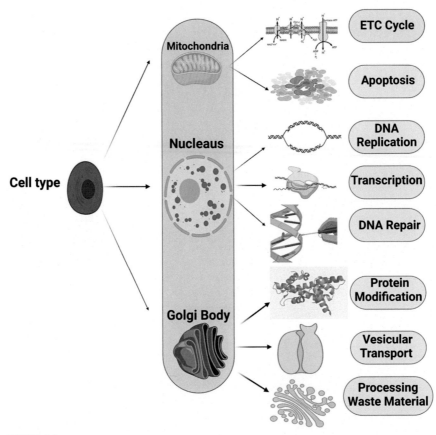

FIGURE 2.2

Hierarchical database model example.

connections and interactions are made. Because the data is more interconnected, accessing it is easier and takes less time.identifies data linkages on a many-to-many scale. Prior to the introduction of relational databases, this was the most commonly used model (Fig. 2.3).

2.3.4 Entity relationship model

In this information base paradigm, links are established by first breaking down the item of interest into its constituent parts, or substances, and then breaking down its attributes, or ascribes. Using connections, several different chemicals are linked together. It is acceptable to design an information base using this model, and that plan would subsequently be able to be converted into tables for use in the social model.In this information base paradigm, links are established by first breaking down the item of interest into its constituent parts, or substances, and then breaking

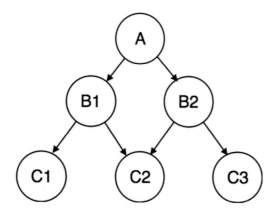

FIGURE 2.3

Network database model example.

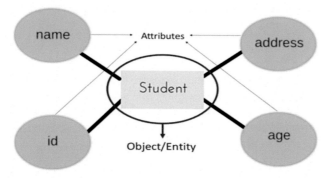

FIGURE 2.4

Entity relationship model example.

down its attributes, or ascribes. Using connections, several different chemicals are linked together. It is acceptable to design an information base using this model, and that plan would subsequently be able to be converted into tables for use in the social model (Fig. 2.4).

2.3.5 Relational database model

The relational database model is now the most well-known information storage type that is accessible. The data in this model are set up in two-dimensional tables, and a common field is used to keep the link between the tables. Two-dimensional tables serve as the primary organizational framework for the information included in this model (basic flat file database). All of the information that may be categorized under a certain heading is saved in the appropriate column of that table. After that, relations are constructed between the several tables, hence the name. The relational database model is now the most well-known information storage type that is accessible. The data in this model are set up in two-dimensional tables, and a common field is used to keep the link between the tables.Two-dimensional tables serve as the primary organizational framework for the information included in this model (basic flat file database). All of the information that may be categorized under a certain heading is saved in the appropriate column of that table. After that, relations are constructed between the several tables, hence the name.

A relational database would be incomplete without a primary key and a foreign key.

- Primary key: It is the identifier column's purpose to store values that are one-of-a-kind (there is no room for duplication), and it is also the column that is used to describe a specific record inside a table in such a way that there is no overlap in the information stored. An example of this might be an accession number, index number, or any of the like.
- Foreign key: It is the column that connects one table to another table that does this (the primary key can also be made the foreign key). Relational databases are distinguished by this one-of-a-kind characteristic.

In 1970, E.F. Codd developed the relational model as a way to make information base management frameworks more independent of a particular application. This was done via the use of a relational database. It is a numerical model that has been defined in terms of predicate reasoning and set hypothesis, and different computer frameworks have used different implementations of it. Centralized server, mid-range, and microcomputer frameworks have all made use of it. In reality, the things that are often referred to as social information bases put into action a model that is an estimate of the numerical model that Codd outlined. Relations, attributes, and regions are the three fundamental concepts that are applied extensively in social information base models. A link may be seen as a table consisting of segments and columns (Fig. 2.5).

2.3.6 Other models

The following is a list of some more models that are not as often used but are applied in certain circumstances:

- **Inverted File Model:** The information itself is used as a key in a query table, and the characteristics of the table are pointers to the location of each instance of a specific content item. The query table is used to organize the information.
- **Dimensional Model:** The dimensional model is a special kind of social model that is used to communicate with the data stored in information distribution centers in such a way that the data may be easily summarized by making use of online scientific handling or OLAP queries.
- **Graph Model:** Graph information bases make it possible to have a far more open structure than an organization data set does; each hub may be connected to another hub.

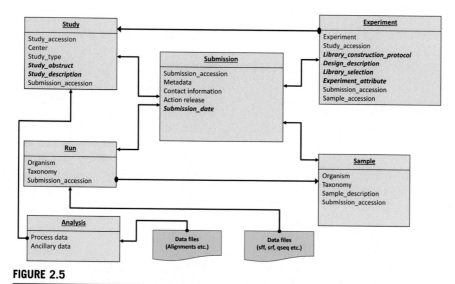

FIGURE 2.5

Relational database explained through the example of SRA (sequence read archive, repository of HTS sequence data available on NCBI).

- **Multivalue Model:** They can store the same path as relational databases, but in addition, they grant a degree of profundity that the relational model can only approximate by using sub-tables. This makes multivalue information bases "knotty" information in the sense that they can store the same path as relational databases.

2.4 Primary nucleic acid databases

The nucleotide database is a collection of different categories derived from a few different sources, such as GenBank, RefSeq, and TPA. The information about the genome, its quality, and the record layout provides the foundation for both biological research and the discovery of new knowledge. Primary nucleotide databases are a kind of biological database that holds data on nucleotide sequences that were obtained directly from researchers doing experiments in university labs, independent laboratories, or sequence centers. GenBank, EMBL, and DDBJ are the three most important databases in terms of the storage and accessibility of crude nucleic acid sequences to both the general public and analysts specifically. They are the repository of all raw nucleotide sequences, which is why they are considered to be the most important nucleotide sequence databases. The National Institute of Health at United States of America hosts the GenBank data repository, which may be accessed through the NCBI website. The DNA Databank of Japan (DDJB) is located in Japan, whereas the European Molecular Biology Laboratory (EMBL) is located in Europe.

To establish the highest level of synchronization possible across the three, each of them will accept submissions of nucleotide sequences, after which they will share fresh and updated data daily. Because they include the original sequencing data, these three databases are considered primary databases.

2.4.1 EMBL

Nucleotide Sequence Collection of the European Molecular Biology Laboratory (EMBL) is a comprehensive collection of primary nucleotide sequences that are maintained by the European Bioinformatics Institute (EBI). There are many different sources of data, some of which include individual scientists, gene sequencing facilities, and patent offices.

2.4.2 GenBank

The GenBank nucleotide sequence database is a collection of all of the publicly available nucleotide sequences and the protein interpretations of those sequences. This database is open to the public and contains annotations. As a crucial part of the NCBI, it is responsible for supplying and maintaining this data collection (INSDC). Get access to the DNA sequences of base pairs that were produced in laboratories all around the globe based on the characteristics of more than 100,000 different species. GenBank has developed into a valuable resource for scientists doing studies in natural environments. This database is now expanding at an exponential pace, as it has done ever since it was first created. The rate at which it is filling up is also exponential.

2.4.3 DDBJ

Its actual location may be found at Japan's National Institute of Genetics (NIG), which is located in the prefecture of Shizuoka. The DDBJ is the only nucleotide sequencing database that is specifically intended for use in Asia. Although Japanese researchers make up the bulk of DDBJ's data pool, the organization also welcomes contributions of sequences from researchers and donors from other nations.

2.5 Primary protein databases

The information included in biological databases may be extensively structured into data sets according to sequence and structure. The succession information bases are important for both the protein groupings and the nucleic acid arrangements, but the structure information bases are only relevant to the proteins themselves. After the insulin protein sequence was made available to the public in 1956, the primary knowledge base was developed in a very short amount of time. Insulin, which was not considered to be the most important protein to sequence, really is. Insulin's structure is shown by just 51 deposits, which are functionally analogous to letter sets in a sentence. These deposits form the sequence.

Around the middle of the 1960s, a nucleic corrosive cluster of yeast tRNA with 77 bases, which are single units of nucleic acids, was discovered for the first time. During this period, research was conducted into the three-dimensional structure of proteins, and in 1972, the now-famous Protein Data Bank was formed as the primary information source on protein structure. Back then, there were just 10 structures in all and currently it has more than 2,00,000 structures diposited.

While the basic information about how proteins are put together was kept at each research facility, in 1986 work began on putting all of this information into one place. This set of information is called the SWISS-PROT protein grouping data set.This data set currently has approximately 70,000 protein sequences from over 5000 model creatures, which is a small portion of every known life form.

Both academic institutions and businesses now have access to these enormous databases containing a wide variety of information sources, making it possible for them to conduct research and investigate the data. These are made available as open-access data in the greater context of the research network through the Internet. These information databases are regularly updated with the addition of new passages.

2.5.1 **PDB**

The Protein Data Bank, often known as PDB, is the only and most widely used library that includes both structural information and pictures of biological macromolecules. The Protein Data Bank (PDB) is the primary information source for many of the derived databases.It is where almost all of the structural bioinformatics studies that have ever been done begin (Joosten et al., 2011).

Files of biological molecules, along with their 3D coordinates, are what are largely kept in the PDB archive as the information that is saved there. These data detail the particles that make up each protein as well as their three-dimensional area in space. The aforementioned records may be accessed in a variety of formats (PDB, mmCIF, XML). The "header" section of a typical PDB-organized document is quite large and contains a lot of text. This section summarizes the proteins, reference data, and nuances of the structural arrangement. This section is followed by the succession and a not-insignificant rundown of the particles and their directions. The exploratory senses that are used to determine these atomic coordinates are also included in this file, which can be found in its entirety.

The RCSB Protein Data Bank adds to the data by making tools and resources for studying and teaching in the fields of molecular biology, structural biology, computational biology, and other related fields.

2.5.2 **SWISS-PROT**

The amino acid sequences that are utilized to connect amino acid sequences to information that is presently kept in the field of life sciences are retrieved from the SWISS-PROT protein database. An overview of the pertinent information is provided for each protein entry by combining the findings of the primary investigations with the characteristics estimated by simulations and other predictions, which may on occasion lead to contradictory conclusions. This is done to provide a comprehensive picture of the situation. This ends up producing an overview of the data that is applicable across disciplines. By establishing direct linkages to a variety of information databases, it is possible to get access to specialized knowledge that is not covered by SWISS-PROT. The annotation of human (the HPI project) and other model living creature passages is the primary emphasis of SWISS-PROT, but it does include explanatory elements for all species. This is done to ensure that sufficient annotation is accessible for individual agent proteins belonging to all protein families. As the High-quality Automated and Manual Annotation of Microbial Proteomes (HAMAP) project has done for species, some explanations may be applied to other families. This permits a more comprehensive understanding of the phenomenon to be gained. To stay current with the most recent findings in the field of logic, protein families and protein groups are subjected to ongoing analyses on a regular basis. By incorporating a growing amount of typically automated explanations, TrEMBL is working toward its eventual goal of covering all protein categories that are not yet handled in SWISS-PROT. This will be accomplished once all protein categories have been covered in SWISS-PROT (Boeckmann et al., 2003).

2.6 **Secondary protein databases**

A secondary database stores information that was either obtained from the original database or another secondary database. The primary database is the source of the information included in the secondary database. Information like the conserved sequence of a protein family, active site residues (derived from multiple sequence alignments of a group of proteins linked functionally or by sequence), and signature sequences are stored in a secondary database (identifying sequence). The Protein Data Bank (PDB) may be broken down into its parts, each of which is referred to as a secondary structural information base. Each alpha protein, each beta protein, and so on all have portions that are arranged in a certain order because of how the protein is structured. In addition, they provide information on the observed optional structural motifs of the protein. Components of the optional information base include SCOP, which was developed at Cambridge University; PROSITE, which was developed at the Swiss Institute of Bioinformatics; CATH, which was developed at University College London; and eMOTIF, which was developed at Stanford. Each of these parts was made by different analysts at their own research centers, where they are also being run.

2.6.1 CATH

CATH is a database that essentially holds a hierarchical categorization of domains of proteins based on how they fold. The domains that are stored in the Protein Data Bank (PDB) are recognized and categorized both manually and via the use of automated computer processes, and they are then placed into CATH. The CATH online interface allows for both simple searches for the categorization as well as thorough searches and downloads of the data (Sillitoe et al., 2019).

- **Class** is generated from the content of the secondary structure, and for more than 90% of the structures that have been saved for proteins, it has been automatically given.
- **Architecture** is a term that is used to define the fundamental alignment of secondary structures, apart from their connectivities (currently assigned manually).
- The **topology** level organizes the structures into groups based on how closely connected they are topologically and how many secondary structures they have.
- **Homologous superfamilies** are groups of proteins that have extremely similar activities and structures to one another.

2.6.2 SCOP

The purpose of the SCOP database is to provide descriptive and holistic details of the structural and evolutionary relationships between proteins whose three-dimensional structure is known and deposited in the Protein Data Bank. This information is sought after to better understand how proteins have evolved (Hubbard et al., 1999). The following are the primary levels of the classification:

- The term **"family"** refers to sets of proteins that are genetically and evolutionarily very closely connected. Current techniques of sequence comparison, including BLAST, PSI-BLAST, and HMMER, can identify their connection in the vast majority of instances.
- A **superfamily** is a group of protein domains that are only loosely connected. Their similarity is typically restricted to common structural characteristics, which, when combined with a conserved architecture of active or binding sites, or comparable processes of oligomerization, imply a likely evolutionary heritage. Their similarities are frequently limited to common structural aspects.

2.6.3 Prostate

Documentation values illustrating protein domains, families, and functional information are included in PROSITE, along with relevant examples and profiles that are used to differentiate between the different types.

PROSITE is complemented with a set of rules known as ProRule, which are reliant on profiles and examples. These rules enhance the ability of profiles and examples to skew results by providing additional information about amino acids that are either functionally or structurally important.

2.7 Composite sequence databases

The necessity to search through a variety of resources is eliminated thanks to the use of a composite sequence database, which combines a broad variety of important data sources. In their inquiry calculations, diverse composite information bases make use of unique fundamental data sets following a variety of principles. In addition to this, several search avenues have been streamlined and compiled for your convenience inside the comprehensive database. The National Center for Biotechnology Information (NCBI), which stands for "National Center for Biotechnology Information," keeps these nucleotide and protein databases on their huge, easily accessible, and redundant array of servers and gives researchers free access to them (Fig. 2.6).

2.7.1 Meta-databases

Metadatabases are knowledge bases that collect information about data sets to produce new data. This information is then used to generate new data. They could put together information from many different sources and present it in a new and more useful way, or they could focus on a certain disease or organism.

Examples are:

- An information discovery service was developed by the University of Antwerp and the Vlaams Instituut Voor Biotechnologie called BioGraph. It is based on the integration of more than 20 different types of databases.
- Information Framework for Neuroscience, hundreds of neuroscience-related resources are integrated into this system
- ConsensusPathDB is a database that integrates information from 12 different databases to provide a comprehensive view of molecular functional interactions.
- Entrez (National Center for Biotechnology Information)

2.8 Genomics and proteomics databases

Genetic information bases, which are also known as online genomic variation stores, may be provided for a single (locus-explicit) or several (generic) traits, or they may be provided expressly for a population or ethnic group (national or ethnic). In either case, the databases may be referred to as "online genomic variation stores." Both of these scenarios are not just possible, but likely. Genomic information bases are critical components of human genome informatics, which has grown exponentially in the postgenomic period as a consequence of a better understanding of the genetic etiology of human diseases and the visible confirmation of various genomic

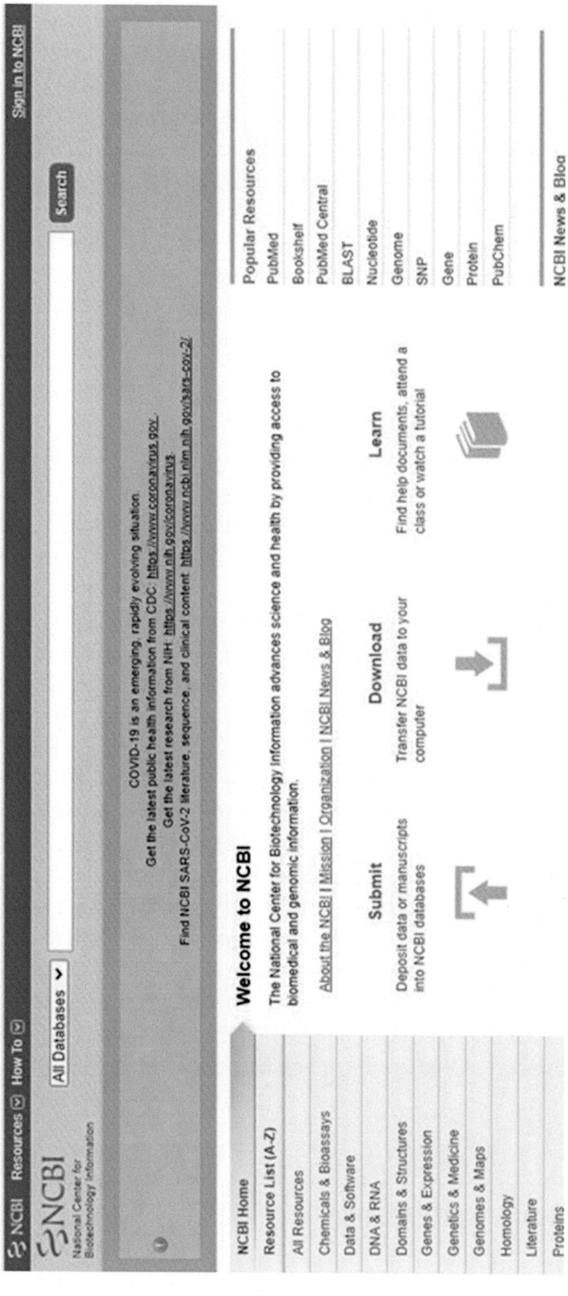

FIGURE 2.6

NCBI is one of the most popular composite databases among researchers hosted by NIH (National Institute of Health), this is the page loaded when we search ncbi.nlm.nih.gov.

variations. In the postgenomic period, the field of genomics informatics has experienced a period of explosive growth. This resulted in an improvement in our understanding of the genetic etiology of human illnesses as well as the visual confirmation of several different genomic polymorphisms. These resources organize the data and variations that were stated before with the hope that, in the future, they will be valuable not only for molecular diagnostics but also for doctors and analysts. To put it another way, the purpose of these resources is to reduce the amount of effort required for consumers to get the information that they need. The proteome databases are a collection of easily searchable, species-specific protein databases that merge publicly available sequence information with material that has been brought up to date via the careful curation of the scientific literature by trained professionals. Access to these databases is provided via the internet. As an example, the TUM serves as the database host for ProteomicsDB (Schmidt et al., 2018).

2.8.1 The search engines for literature

Internet access enables rapid access to a large quantity of clinical writing, including diaries, databases, word references, course readings, files, and electronic diaries. This, in turn, enables access to more variable, personalized, and effective instructional options. An online internet searcher is a piece of software or hardware designed to search for information on the World Wide Web. This information may take the form of website pages, photos, data, or other types of records. Web search tools for the web-based pursuit of clinical writing include Google, Google researcher, Yahoo internet searcher, and so forth, and information bases include MEDLINE, PubMed, MEDLARS, and so forth. Business web assets (Medscape, MedConnect, and MedicineNet) add to the rundown of asset information bases, giving a portion of their substance to open access. Some online libraries, such as the Medical Framework and the Emory Libraries, have been established as meta-destinations, providing important linkages to various wellness resources located all over the world. The availability of specific websites about dermatology, such as DermIs, DermNet, and Genamics Jornalseek, is a valuable addition to the list of electronic resources, which is always expanding. When searching for a certain category of information, a scientist has to keep in mind the benefits and drawbacks of the web crawler or data collection that they are using. In the field of medicine, information about the types of writing and levels of detail that are available, the user interface (UI), how easy it is to get started, how reliable the content is, and the time period that is covered makes it possible to get the most out of these resources.

2.9 Miscellaneous databases

2.9.1 Humans

Over the last 5 years, the Human Genome Project has had a significant amount of effect on the field of genetics research. This influence will soon be felt across the entire biological and medical professions. It won't be long now. After describing them for centuries, we are now on the cusp of having a perfect comprehension of the cellular processes that take place within the human body. This is despite the fact that we have been describing them for millennia. Linkage analysis in families using limited sets of genetic markers was a technique that was necessary up until the late 1980s to establish the origin of hereditary disorders. This approach was time-consuming and required the use of genetic markers. At the end of that decade, a more all-encompassing strategy was proposed. It was called the Human Genome Project, and it included the mapping of all 80,000–100,000 of our genes as well as the decoding of our entire DNA sequence, which is 3 billion base pairs in length. This project was intended to be completed by the end of that decade. The first and most notable result of this program has been a spectacular acceleration in the process of determining the factors that contribute to inherited diseases. This initiative has considerably accelerated the development and distribution of advanced DNA technologies. Ten years have passed since the project's first conception. It has been shown that a malfunctioning gene or genes are to blame for the bulk of common genetic illnesses (150–200), in addition to a considerable number of rare genetic diseases (600–800). The majority of the disease-causing mutations have been uncovered, which has led to a huge improvement in the diagnostic capabilities of the condition. New genetic pathways have come to light as a result of these investigations, which are still ongoing. Genomic imprinting, expansion of triplet-repeat sequences, and defects in DNA repair are examples of some of these processes. This research, in turn, has led to breakthroughs in diagnostic techniques as well as the discovery of more disease-causing genes. The study of the so-called "genotype-phenotype correlation," which makes it possible for a more in-depth examination of the relationship between fundamental molecular flaws and functional disturbances of processes in cells, organs, and the organism as a whole, has also been made possible as a result of advancements made on a global scale. Because there are numerous stages in the chain of events that link cause and effect in the cell, as well as a complex web of interactions between various genes and other components of the environment, it is sometimes exceedingly difficult to establish these correlations. This is because genes and other elements of the environment interact in a complex web.

The Genome Database is a repository for data about human genes, clones, STSs, polymorphisms, and maps. This data may be accessed by the general public. The entries in the GDB are very well related to one another, to citations from the scientific literature, and to entries in other databases, including the sequence databases (Letovsky et al., 1998), OMIM, and the Mouse Genome Database. GDB is continuously receiving fresh mapping data from a variety of sources, including big genome centers as well as smaller mapping efforts. The database may be searched using a variety of methods, ranging from simple keyword searches to more involved queries. Over the last year, a significant number of brand-new capabilities have been included. For instance, Comprehensive Maps, which are integrated maps of the human genome and are now in the process of being created, are being used to facilitate positional searches and visual presentations. Printing maps and displaying ad hoc query results in a graphical style are two new features that have been added to the

GDB map viewer, often known as Mapview. The HUGO Nomenclature Committee is continuing to collaborate with GDB to maintain a record of the proposed and official gene symbols, in addition to the data associated with them. Because genome research is moving away from mapping and toward sequencing and functional analysis, the GDB schema is getting bigger.

2.9.2 Animals

Agriculture focused on animals will have to adapt swiftly if it is to be successful in meeting the food requirements of the future. The study of an organism's entire gene sequence is known as genomics, and it is a significant factor in the development of novel agricultural practises and technologies. Healthier animals that develop quicker, are less likely to become ill, and are better equipped to manage stressful or changing surroundings may be the result of using genomic information to improve how animals are bred. This may lead to the discovery of new animal breeds. Eliminating these issues would not only enhance the health of the general population but also cut expenses and losses for farmers. The satisfaction of clients may be increased by producing higher-quality goods. In addition, genomics may shed light on novel approaches to managing livestock production systems that are both more effective and friendlier to the natural environment. However, research into animal genomes is laborious, time-consuming, and costly. It might be too costly for individual researchers or smaller institutions to get the necessary tools and knowledge about genes. Because they do not know much about the latest technology, some scientists could be reluctant to employ them. Researchers will need to work together to make progress in animal genome research and find new ways to use what they know.

Experiments led to the discovery of the sizes of the genomes of over 6000 different animal species, and this information is compiled in a database called the Animal Genome Size Database. Each entry contains information on the taxonomy of the species, common names, how and from what tissues the size of the species' genome was estimated, and other relevant details. The database also includes connections to other locations on the internet where users may get images and further information about the species of their choice.

2.9.3 Fungi

To keep using fungi for the good of people, though, it's important to know how these organisms interact in both natural and artificial communities. People will soon be able to collect samples from different locations to investigate the possibility of complex fungus metagenomes. This will be an essential component of using fungus for the goals of industrial production, energy production, and climate management. However, unless we have well-defined reference data for fungal genomes, we won't be able to do a thorough analysis of these data.

An international research team is collaborating with the Joint Genome Institute of the Department of Energy on a project that will last for 5 years and will sequence the genomes of 1000 different types of fungi that are found across the Fungal Tree of Life. This will allow the team to learn more about the diversity of fungi. It is intended that the Fungal Tree of Life will be completed by sequencing at least two reference genomes from among the more than 500 families of fungi that are currently known. To do this, the main goal of this project is to collect data that can be used as a starting point for future studies on how plants and microbes interact, how microbes release and absorb greenhouse gases, and how metagenomes from the environment are sequenced.

A resource for functional genomics about pan-fungal genomes may be found inside the FungiDB database. It was put together by the Eukaryotic Pathogen Bioinformatics Resource Center, which contributed to its creation. Both the layout of FungiDB and its user interface are reminiscent of those seen in EuPathDB. This implies that complex and integrated searches may be done with a graphical method that is easy to comprehend. The most recent iteration of the FungiDB database contains the genomic sequences and annotations of 18 distinct fungus species that are representative of a wide range of fungal groups. The Basidiomycota orders Pucciniomycetes and Tremellomycetes, the Ascomycota classes Eurotiomycetes, Sordariomycetes, and Saccharomycetes, and the Mucormycotina lineage, which is the most fundamental "Zygomycete" lineage, are all included in this category. More information is available on the cell cycle microarray, the hyphal growth RNA sequence, and the yeast two-hybrid interaction in the FungiDB database.

2.9.4 Microorganisms

In our world, the group of creatures with just a single cell, known as microbes, is without a doubt the most abundant and diverse. There are approximately 12,000 known species, but there are likely millions more waiting to be discovered on our planet. Bacteria are capable of surviving practically every environment on Earth, including those that do not seem to be favorable to the existence of life. They have been seen as low as seven miles under the surface of the water and as high as 40 miles in the sky. There are a great number of types of bacteria that can survive in extreme environments such as intense heat, cold, and salt. The genomic sequences of microorganisms give a wide diversity of strains with varying degrees of quality and sampling intensity. Several significant diseases affect humans among them, as well as species that are interesting for reasons other than those related to medicine, such as biodiversity, epidemiology, and ecology. We have learned a great deal about evolution, microbial biology and ecology, and the existence of microbes from studying many different kinds of microbes, such as obligate intracellular parasites, symbionts, free-living bacteria, hyperthermophiles, psychrophiles, and aquatic and terrestrial microorganisms. The archeal genome of *Candidatus Parvarchaeum acidiphilum* was found to range in size from 45 kb to the largest draught assembly of *Mastigocoleus testarum* and the largest whole genome (14.7 mb) of *Sorangiumcellulosum*, an alkaline-adapted epiphyte maker. This discovery was made through research on mine drainage metagenomes.

On a workstation, the MBGD system allows for comparative and analytical examination of completely sequenced microbial genomes. The primary objective of the MBGD method is to generate an orthologous gene categorization table. This is accomplished by using all-against-all similarity correlations that have previously been computed between genes located in different genomes. By providing a list of organisms and parameters, users of MBGD will be able to generate their very own classification tables thanks to the incorporation of an automatic classification algorithm within the software. When the user is interested in creatures that belong to the same taxonomic category, this function is extremely beneficial since it allows them to narrow their search. The categorization table that was created has been kept in the database, where it may be seen in conjunction with data from individual genomes and information about the degree to which individual genomes are similar to those of other organisms.

2.9.5 Plant and crop genomic database

PlantGDB is a database that contains information on the molecular sequences of all plant species that have had major attempts made to sequence them. EST sequences are organized in the database into groupings referred to as "contigs." These contigs are assumed to represent individual genes. Contigs are assigned labels and, in cases where it's feasible, connected to the sections of genomic DNA to which they belong. The individual components of the genome sequence are assembled using the same method. The website known as PlantGDB aims to provide a method for locating groups of genes that are present in all plant species or that are exclusive to certain plant species alone. This is accomplished by integrating a variety of different bioinformatics tools, each of which makes it simpler to predict genes and compare them across different species. You can see the genomes of different species using PlantGDB that have undergone large-scale genome sequencing initiatives. It does this by putting together all of the EST and cDNA evidence for the gene models that are already available.

2.9.6 Organelle database

The online interface for the relational database known as Organelle DB may be found at the following address: http://organelledb.lsi.umich.edu. It includes a rundown of the key protein complexes as well as the proteins that may be found in organelles. Since its launch in 2004, Organelle DB has had a 20% expansion in size. It now contains over 30,000 proteins from 138 different eukaryotic species. The information provided by Organelle DB includes the location of each protein in the cell, the main sequence of the protein, and, if it is available, an in-depth description of the function that the protein serves. Every entry in Organelle DB has been annotated with words taken from the controlled vocabulary that is maintained by the Gene Ontology collaboration. The facts associated with protein localization are intrinsically visual, and Organelle DB contains a massive collection of photos related to biology. It includes 1500 micrographs of yeast cells that have been labeled to show the proteins.

2.9.7 Pathway databases

Pathway databases enable the compilation of lists of proteins and their associated functions, as well as the integration of such lists into networks that depict the functioning of an organism. The Reactome Knowledgebase is used as an example to show how a reaction space is built from manually curated experimental data, how semi-automated extensions of these manual annotations can be used to infer annotations for a large portion of a species' proteins, and how networks of functional annotations can be used to infer pathway relationships between different proteins that have been linked to disease risk by genome-wide studies.

References

Bateman, A., Martin, M.J., O'Donovan, C., Magrane, M., Alpi, E., Antunes, R., Bely, B., Bingley, M., Bonilla, C., Britto, R., Bursteinas, B., Bye-AJee, H., Cowley, A., da Silva, A., de Giorgi, M., Dogan, T., Fazzini, F., Castro, L.G., Figueira, L., Garmiri, P., Georghiou, G., Gonzalez, D., Hatton-Ellis, E., Li, W., Liu, W., Lopez, R., Luo, J., Lussi, Y., MacDougall, A., Nightingale, A., Palka, B., Pichler, K., Poggioli, D., Pundir, S., Pureza, L., Qi, G., Rosanoff, S., Saidi, R., Sawford, T., Shypitsyna, A., Speretta, E., Turner, E., Tyagi, N., Volynkin, V., Wardell, T., Warner, K., Watkins, X., Zaru, R., Zellner, H., Xenarios, I., Bougueleret, L., Bridge, A., Poux, S., Redaschi, N., Aimo, L., ArgoudPuy, G., Auchincloss, A., Axelsen, K., Bansal, P., Baratin, D., Blatter, M.C., Boeckmann, B., Bolleman, J., Boutet, E., Breuza, L., Casal-Casas, C., de Castro, E., Coudert, E., Cuche, B., Doche, M., Dornevil, D., Duvaud, S., Estreicher, A., Famiglietti, L., Feuermann, M., Gasteiger, E., Gehant, S., Gerritsen, V., Gos, A., Gruaz-Gumowski, N., Hinz, U., Hulo, C., Jungo, F., Keller, G., Lara, V., Lemercier, P., Lieberherr, D., Lombardot, T., Martin, X., Masson, P., Morgat, A., Neto, T., Nouspikel, N., Paesano, S., Pedruzzi, I., Pilbout, S., Pozzato, M., Pruess, M., Rivoire, C., Roechert, B., Schneider, M., Sigrist, C., Sonesson, K., Staehli, S., Stutz, A., Sundaram, S., Tognolli, M., Verbregue, L., Veuthey, A.L., Wu, C.H., Arighi, C.N., Arminski, L., Chen, C., Chen, Y., Garavelli, J.S., Huang, H., Laiho, K., McGarvey, P., Natale, D.A., Ross, K., Vinayaka, C.R., Wang, Q., Wang, Y., Yeh, L.S., Zhang, J., 2017. UniProt: the universal protein knowledgebase. Nucleic Acids Res. 45, D158–D169. https://doi.org/10.1093/NAR/GKW1099.
Bilofsky, H.S., Christian, B., 1988. The GenBank ® genetic sequence data bank. Nucleic Acids Res. 16, 1861–1863. https://doi.org/10.1093/NAR/16.5.1861.
Boeckmann, B., Bairoch, A., Apweiler, R., Blatter, M.C., Estreicher, A., Gasteiger, E., Martin, M.J., Michoud, K., O'Donovan, C., Phan, I., Pilbout, S., Schneider, M., 2003.

The SWISS-PROT protein knowledgebase and its supplement TrEMBL in 2003. Nucleic Acids Res. 31, 365–370. https://doi.org/10.1093/NAR/GKG095.

Gligorijević, V., Pržulj, N., 2015. Methods for biological data integration: perspectives and challenges. J. R. Soc. Interface 12. https://doi.org/10.1098/RSIF.2015.0571.

Hubbard, T.J.P., Ailey, B., Brenner, S.E., Murzin, A.G., Chothia, C., 1999. SCOP: a structural classification of proteins database. Nucleic Acids Res. 27, 254–256. https://doi.org/10.1093/NAR/27.1.254.

Joosten, R.P., te Beek, T.A.H., Krieger, E., Hekkelman, M.L., Hooft, R.W.W., Schneider, R., Sander, C., Vriend, G., 2011. A series of PDB related databases for everyday needs. Nucleic Acids Res. 39, D411–D419. https://doi.org/10.1093/NAR/GKQ1105.

Kanz, C., Aldebert, P., Althorpe, N., Baker, W., Baldwin, A., Bates, K., Browne, P., van den Broek, A., Castro, M., Cochrane, G., Duggan, K., Eberhardt, R., Faruque, N., Gamble, J., Garcia Diez, F., Harte, N., Kulikova, T., Lin, Q., Lombard, V., Lopez, R., Mancuso, R., McHale, M., Nardone, F., Silventoinen, V., Sobhany, S., Stoehr, P., Tuli, M.A., Tzouvara, K., Vaughan, R., Wu, D., Zhu, W., Apweiler, R., 2005. The EMBL nucleotide sequence database. Nucleic Acids Res. 33, D29–D33. https://doi.org/10.1093/NAR/GKI098.

Letovsky, S.I., Cottingham, R.W., Porter, C.J., Li, P.W.D., 1998. GDB: the human genome database. Nucleic Acids Res. 26, 94–99. https://doi.org/10.1093/NAR/26.1.94.

Okubo, K., Sugawara, H., Gojobori, T., Tateno, Y., 2006. DDBJ in preparation for overview of research activities behind data submissions. Nucleic Acids Res. 34, D6–D9. https://doi.org/10.1093/NAR/GKJ111.

Rother, K., Michalsky, E., Leser, U., 2005. How well are protein structures annotated in secondary databases? Proteins Struct. Funct. Bioinf. 60, 571–576. https://doi.org/10.1002/PROT.20520.

Schmidt, T., Samaras, P., Frejno, M., Gessulat, S., Barnert, M., Kienegger, H., Krcmar, H., Schlegl, J., Ehrlich, H.C., Aiche, S., Kuster, B., Wilhelm, M., 2018. ProteomicsDB. Nucleic Acids Res. 46, D1271–D1281. https://doi.org/10.1093/NAR/GKX1029.

Sillitoe, I., Dawson, N., Lewis, T.E., Das, S., Lees, J.G., Ashford, P., Tolulope, A., Scholes, H.M., Senatorov, I., Bujan, A., Ceballos Rodriguez-Conde, F., Dowling, B., Thornton, J., Orengo, C.A., 2019. CATH: expanding the horizons of structure-based functional annotations for genome sequences. Nucleic Acids Res. 47, D280–D284. https://doi.org/10.1093/NAR/GKY1097.

Statistical methods in bioinformatics

3

3.1 Introduction

Statistics is the study of collecting, mining, organizing, analyzing, and presenting data in a manner that draws a meaningful conclusion about the set of data. This allows us to utilize smaller sets of samples for making precise and wise interpretations about a larger population. When statistics are incorporated into bioinformatics, it aids in an easier and much faster analysis of biological data. The main objective of statistics and bioinformatics is to assist researchers in acquiring equitable solutions to various biological problems using computational data analysis. Statisticians bring not only a distinct viewpoint but also a set of skills to this approach, putting them at the center of computational research. Understanding fluctuations and uncertainties in quantification is among the fundamental characteristics that distinguish a statistician from other mathematicians. These are critical factors to take into consideration when developing reliable techniques for biological discoveries and verification, particularly when dealing with complex, high-dimensional data like that found in genomic information. The decisions made when designing samples have big effects on the analysis that comes after. Decision making can influence the multistep processing techniques while reducing data loss caused by extraction methods and may be responsible for propagating the errors in the entire process, so accurate decision making is prioritized using the standard protocols to reduce the errors. In the next few sections, we will discuss about some statistical methods used to make sense of biological data.

3.2 Statistics at the interface of bioinformatics

The word "bioinformatics" refers to a multidisciplinary area in which computational researchers, mathematical data analysts, systems biologists, and statisticians investigate multiple dimensions of biological data, including its storage, retrieval, organization, and detailed evaluation. Due to the numerous obstacles presented by this complicated discipline, bioinformatics must be multidisciplinary. It is not possible for a single researcher to possess the medical, biological, computational, data repository management, mathematical analysis, and statistical skills and knowledge

All About Bioinformatics. https://doi.org/10.1016/B978-0-443-15250-4.00009-5

necessary to locate and validate the large amount of scientific information contained in the outputs of these technologies. In bioinformatics, some of the most important things to study are sequence alignment, gene discovery, gene assembly, protein structure alignment and its predictions, gene expression estimation, protein-protein interaction, and evolutionary modeling. Since the bulk of biological information is not stored in a database reduced to a single, flat record, data extraction from structured information is especially significant for bioinformatics application domains. Bioinformatics databases are set up with elements that are connected to each other and to other elements by links that describe a complex structure. The unique viewpoint and skill set of statisticians position them at the core of this process. Awareness of variability and unpredictability in assessment is one of the major characteristics that set statisticians apart from all other quantitative disciplines. These are critical factors for constructing reproducible techniques for biomedical exploration and testing, particularly when dealing with complex, high-dimensional data such as those observed in genomics. Statisticians are "data scientists" who know the affects of sampling design decisions on later stage analysis, how biases can spread through multi-step processing methods, and how information can be lost if feature extraction methods are too simple. They are experts in inferential reasoning, which enabled them to recognize the significance of multiple statistical revisions to avoid reporting spurious results as discoveries and to structure algorithms to hunt high-dimensional spaces and construct predictive statistical models while obtaining precise measurements of accuracy (Fig. 3.1).

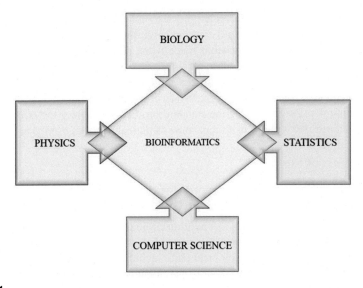

FIG. 3.1

Bioinformatics, an interdisciplinary field of science, aims at using concepts from major branches like statistics, biology, computer science and physics.

There are two types of statistical methods used in analyzing biological data: first one is descriptive statistics while the other one is inferential statistics.

For describing data from a representative cohort, descriptive statistics utilize the mean and standard deviation. It is a type of statistical analysis that looks at a population using math, graphics, and tables. Descriptive statistics assist in portraying data in a more meaningful fashion, allowing for easier data interpretation.

Inferential statistics views information as a subgroup of a given sample, using datasets from that sample to draw conclusions and make calculations about that group. Inferential statistics allows one to establish generalizations regarding a group using sample data. Hypothesis testing, confidence intervals, and linear as well as multiple regression analysis are the most frequent inferential statistical approaches.

3.3 Measures of central tendency

Before going on to statistical analysis, it is required to review the raw data. Then, the two most significant sample statistics that can be derived from a dataset are a measure of the sample distribution's central tendency and the data's dispersion around this central tendency. These descriptive statistics are necessary for inferential statistical analysis. In simple terms, "central tendency" is a technique for describing the dataset's core. It is also known as a distribution center or site. Various measures of central tendency strive to define what is variably referred to as the typical, normal, expected, or mean value of a dataset. The mean, the mode, and the median are widely utilized for the majority of data types. The mean is one of the most common measures of central tendency, and most people think it's the best. However, in some cases, either the median or the mode is preferable. Whenever there are some extreme values in the given data, the median seems to be the recommended measure for the central tendency (Smucker et al., 2018).

3.3.1 Mean

Statistically, a dataset's mean is a single number that indicates the midpoint or average value of all the data points. One way to gauge central tendency is through the use of the mean, which is sometimes known as the arithmetic average. For determining the mean, the sum of all data elements is multiplied by the number of times it has been observed. This is the most common and widely used statistic. It can be used to characterize a population sample containing a singular value representing the data's center.

$$\text{Mean} = \frac{\text{Sum of all the observations}}{\text{Total Number of observations}}$$

Ideally, the mean represents the region in which the majority of the values in a distribution fall. The mean does not, however, necessarily locate the center of the dataset. It is susceptible to extreme data and can be skewed. This issue emerges when outliers have a considerable influence on the mean. As the skewness of the distribution increases, the average moves further from the center. In some instances, the mean might be deceptive since it may not be close to the most common values. Therefore, it is optimal to use the mean to identify the central tendency in a symmetric distribution. For skewed distributions, it is frequently preferable to utilize the median, which locates the center using a different approach. The mean doesn't tell us anything about how different a distribution is, so we need to look at the standard deviation to figure out how different it is.

3.3.2 Median

It's a straightforward metric for determining central tendency. For information categorized at an ordinal level, that is the most appropriate indicator of average. The median in statistics is the value that divides a ranked group of data entries in half and corresponds to the 50th percentile of the dataset. It is exactly in the center of the dataset, with half the numbers below and half above. Locating the median is simple. Sort the values in the dataset from smallest to greatest. Then, find the value with the same number of data entries above and below it. The middle value is determined differently depending on whether your data has an even or odd number of values. This section shows you how to figure out the median for data that is grouped or ungrouped.

3.3.2.1 Median for the grouped data

The following formula is used to find the median when the data points are in odd number:

$$\text{Median} = \left(\frac{N+1}{2}\right)^{th} \text{term}$$

The following is the formula for calculating the median for even number of data points:

$$\text{Median} = \frac{\frac{N^{th}}{2}\text{term} + \left(\frac{N}{2}+1\right)^{th}\text{term}}{2}$$

Here, "N" represents the entire number of elements/observations in the sample dataset and "th" indicates the $(N)th$ number.

3.3.2.2 Median for the ungrouped data

If the data is well grouped then we use the following formula to find the median:

$$\text{Median} = l + \left(\frac{\frac{N}{2} - cf}{f}\right) \times h$$

here, "*l*" is the median class lower limit, "*cf*" represents the cumulative frequency of class previous to the dataset's median class. "*f*" signifies the frequency of median class and "*h*" represents the size of class.

One can gain an impression of a dataset's dispersal by contrasting its median with the mean. The set of data is almost uniformly dispersed across lowest and highest values whenever the mean and median value are identical. When the sample distribution is reported to be skewed, the median is a more accurate estimate of central tendency than the mean since it is less vulnerable to outliers. Extreme values drive the mean distant from the distribution's center, rendering it potentially deceptive.

3.3.3 Mode

Similar to the mean and median, the mode is a statistical method to calculate the central tendency. The mode is by far the most prevalent value in a data set, i.e., the characteristic or value that occurs most frequently in any particular dataset. To put it another way, the mode of a set of numerical data is the variable occurring with the greatest frequency in that dataset. It is an averaged metric that can be applied to nominal variables. There may be more than one mode in a dataset, which is different from other measures of central tendency. A data set can be bimodal (having two modes), trimodal (having three modes), multimodal (having >3 modes), or no-model (having no modes). One of its best features is that it can be used for any kind of data, unlike the mean and the median, which can't be determined for nominal data. In the case of normally distributed data, its mode has the same numeric values as observed in the mean and median, although it can be quite distinct in severely skewed patterns. Mode is unaffected by the most and least extreme numbers in quantitative datasets. So, it can give insights into almost any set of data values, no matter how the data is organized. However, the statistical metric is not without its own limits. For example, it cannot be dealt with mathematically any further. Therefore, it cannot be utilized for a more comprehensive study. In addition, because mode is not based on all values in the dataset, it is challenging to draw inferences about the dataset using mode alone (Krzywinski and Altman, 2014a).

3.3.4 Percentiles, quartiles and interquartile range

Percentiles are remarkably versatile and can be used to produce a relative ranking, to divide a dataset into equal proportions, to determine a distribution's central tendency, and to assess its dispersion. Also, quartiles are a means of describing the distribution of data. Every collection of "*X*" intervals that create boxes having an identical number of observations is referred to as a "quantile" of that variable "*X*" The audience is probably familiar with percentiles, which are boxes comprising 1% of information. Another popular method is the quartile, which uses "4*X*" intervals for defining boxes holding 25% of the information. Whenever distributions contain skewness or excessive kurtosis, the interquartile range (IQR) is usually used instead of the standard deviation since the standard deviation cannot offer dependably decipherable information regarding non-Gaussian distributions. The median represents the 50th percentile. This value halves a given dataset. Half of the results fall under the 50th percentile, while the other half exceed it. Based on percentiles, quartiles are the numbers that split your data into quarters. The 25th percentile corresponds to the first quartile, commonly known as Q1 or the lower quartile. One-fourth of the scores fall below this threshold, while three-quarters exceed it. The median, often known as the second quartile or Q2, is just the value of the 50th percentile. Half of the scores are significantly higher than average, while the other half are significantly lower. The 75th percentile corresponds to the third quartile, commonly known as Q3 or the upper quartile. One-quarter of the scores exceed this number, while the remaining three-quarters are below. In descriptive analysis, the interquartile range is often referred to as the "mid-spread," the "mid 50%," or the "H-spread." It is a statistical dispersion indicator dependent on the division of a data frame across quartiles. The IQR is defined as the gap between the 75th and 25th percentiles, establishing a region of "X" that encompasses the central 50% of distributions, but is not always centered around the average as well as the median (Fig. 3.2) (Krzywinski and Altman, 2014b).

$$IQR = Q3 - Q1$$

The q-q plot, also known as the "quantile-quantile plot," is a method for determining whether two sets of data originate from the same population with the same distribution (Fig. 3.2). By plotting the quantiles of two probability distributions

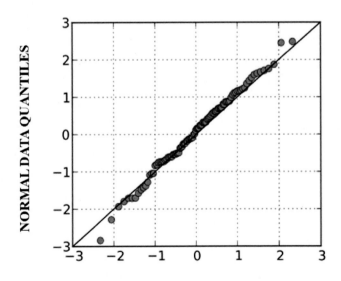

FIG. 3.2

A normal quantile-quantile plot (QQ-Plot).

against each other, Q-Q (quantile-quantile) plots are a very important way to compare and analyze them visually. If the two population distributions we are trying to compare are exactly the same, then the observations on the Q-Q plot will flawlessly continue lying on a straight line following "$y = x$." Q-Q plots are utilized to compare the distributional shapes of two groups. They illustrate how placement, size, and skewness are similar or dissimilar between the two distributions. Q-Q plots can be used to compare data sets or models of possible outcomes.

3.4 Skewness and kurtosis

When sample data considerably deviates from the norm, skewness quantifies the ensuing imbalance. Sometimes, the normal distribution appears to be skewed to one side. This is because the possibility of data being greater than or less than the mean is greater, resulting in an asymmetrical distribution (Fig. 3.3) (Royston, 1992).

As we know, a normal distribution takes the form of a bell curve, any deviation from this normal bell curve leads to skewness, suggesting that the data are not equally distributed. The asymmetry may take two forms:

1. **Positively Skewed:** In a distribution with a positive skew, the values are more packed on the right side, while the left tail is more dispersed. Consequently, the statistical results are left-skewed. Therefore, the mean, the median, and the

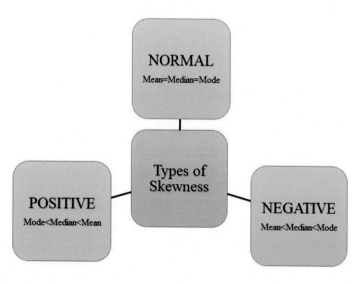

FIG. 3.3

Three types of skewness: Normal where mean, median and mode, all are equal; Positive where mean is greater than median and mode; Negative where mode is greater than mean and median.

mode are always positive. In this type of distribution, the mean is bigger than the median and the mode.

2. **Negatively Skewed:** The right side of the distribution has a higher density of sample points when it is negatively skewed. The average, median, and mode shift to the right as a result. Therefore, these numbers are always negative. In this distribution, the mode exceeds the median while falling short of the mean (Fig. 3.4).

Kurtosis, similar to skewness and is applied to discover data sample outliers. It provides the proportion of outliers present in total. The data may have heavier tails and a flatter peak, comparable to when a distribution is punched or squashed. It is referred to as negative kurtosis (polykurtic). Positive kurtosis indicates a distribution in which the top curve is steeper, indicating that the sample distribution is being pulled upwards (leptokurtic). A normal kurtosis (mesokurtosis) indicates a normal distribution (Fig. 3.5).

The expected value of kurtosis is 3. There is symmetry in this distribution. Positive kurtosis is present if the kurtosis value is greater than three. In this instance, the range of the kurtosis value is between 1 and infinity. A kurtosis of fewer than three also denotes a negative kurtosis. A negative kurtosis has a value range of 2 to infinity. The peak gets higher as the kurtosis value increases (Fig. 3.6).

The typical distributions of variables in everyday life are seldom flawless. Both the skewness and kurtosis coefficients show how much a dispersion is different from data that is normally distributed. Thus, we may say that skewness and kurtosis are used to define the width and height of the normal distribution. The data's horizontal drag is represented by skewness. Kurtosis is used to calculate the vertical drag or the height of the peak, as it reflects how dispersed the data is.

3.5 Variability and its measures

Variability is a statistical method that usually represents the dispersion within a given dataset. A measure of central tendency provides the mean, whereas a measure

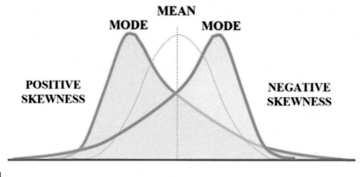

FIG. 3.4

A schematic diagram showing positively skewed and negatively skewed distribution.

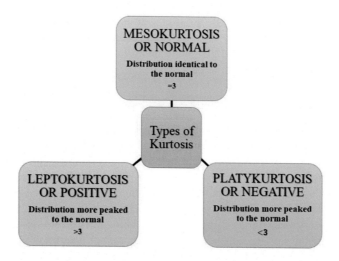

FIG. 3.5

Three types of kurtoses: Normal where distribution is identical to normal; Positive where distribution is more peaked toward the normal; Negative where distribution is less peaked toward the normal.

of variability specifies the proportion of additional data points that are likely to deviate from the mean. Within the context of a value distribution, variability is taken into account. A low dispersion suggests that the sampled data points are firmly grouped around the distribution's center. The greater the dispersion, the further apart they are likely to fall (Hazra and Gogtay, 2016).

Analysts usually use the mean to talk about the center of a group or even a whole system. Since the mean is important, individuals frequently react more to variability. When the variability of a distribution is minimal, the values in a dataset seem to be more stable. However, whenever the variability is increased, the data points become

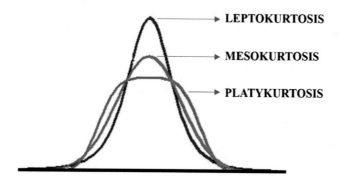

FIG. 3.6

A schematic diagram of different types of kurtoses representing their peaks.

more different and the likelihood of extreme results increases. Therefore, under-standing variability enables you to comprehend the chance of odd occurrences. In the sections that follow, we will give an in-depth look at some of the measures of variability used to understand biological data.

3.5.1 Variance

The prediction of a randomized variable's squared deviation from its mean is called its variance. It determines how distant a group of figures or information are from its mean. The variance compares each value to the mean in order to include all values in the computation. A huge variance shows that the sample units are far off from its mean and each other. To generate this statistic, the squared discrepancies between the data points and the mean are computed, added, and then divided by the total number of observations in the sample dataset. There are two formulas to compute the variance of the sample, depending on whether we are determining the variance for the whole population or using a subset to predict the sample variance of the entire population (Dakhale et al., 2012).

To assess the variation of the whole population, following formula of *Population Variance* is used:

$$\sigma^2 = \frac{\sum (X - \mu)^2}{N}$$

here, "σ^2" represents the variance's population parameter, "μ" represents the population parameter for the population mean, and "N" denotes the number of data points that should represent the overall population.

While, to find the variance of the population using a small subset of population, following formula of *Sample Variance* is used:

$$s^2 = \frac{\sum (X - M)^2}{N - 1}$$

where, "s^2" is the variance for the sample taken from the population and "M" denotes the sample's mean. The inclusion of "$N - 1$" in the denominator compensates for the propensity of samples to underestimate the population's variance.

3.5.2 Standard deviation

The standard deviation shows how far each data point is from the population mean on average. It goes down when the values in a dataset are closer together, indicating a lower dispersion of data values. On the other hand, when there are more differences between the values, the standard deviation goes up because the standard deviation goes up, signifying a higher dispersion in the data values. The standard deviation uses the data's original units, which makes it easier to understand. So, the standard deviation is the most common way to measure how much something can change within a population as well as for a specific subset of that population. The standard deviation is determined by taking the square root of the sample variance, which is a metric that represents the dispersion or variability of a sample in relation to its mean, which represents how far a set's measurements deviate from its average/expected value.

To find the variance of the whole population, following formula of *Population Standard deviation* is used:

$$\sigma = \sqrt{\frac{\sum (X - \mu)^2}{N}}$$

here, "σ" stands for the population standard deviation. "X" signifies the values present in the data distribution, "μ" signifies the population mean while "N" stands for the total number of observations in the population sample.

To find the variance of a sample drawn out of a population, following formula of *Sample Standard deviation* is used:

$$s = \sqrt{\frac{\sum (X - \overline{X})^2}{N - 1}}$$

where, "s" signifies the standard deviation for the sample drawn out of a particular population and "$\overline{X}$" stands for the mean of the sample.

Remember that the square units are used to measure variance, and as a result, the square root helps to restore the value to its native units. Standard deviation is used as a parameter for the whole population and is shown by the symbol "σ", while as an estimate from a sample, however, it is shown by the symbol "s". To figure out the standard deviation, figure out the variance as shown above, and then take the square root of the result. The standard deviation is a number that is like the mean absolute deviation and is considered the most accurate indicator of variation. Both use the original units of the data and compare the data values to the mean to find out how much they vary. But there are differences.

3.5.3 Standard error

Standard error, sometimes known as "standard error of the mean," is an inferential measure that depicts, in simple terms, the degree to which the data from a sample are representative of the total population. For instance, if you conduct a survey of Delhi residents, you will acquire a sample of data that is representative of a portion of Delhi residents. Different types of the same population might thus have different results, so it is essential to evaluate the significance of your findings. The standard error represents the distance between two means when comparing the mean of the sample data to the mean of the complete population on a distribution. How much would the sample mean change if the same study was conducted on a different group of Delhi, India residents? To figure out the standard error, divide the standard deviation by the square root of the sample size "N" for the whole population.

$$SE = \frac{\sigma}{\sqrt{N}}$$

In statistics, sample information is utilized to comprehend larger populations. Probability sampling, in which pieces of a sample are picked at random, enables the collection of data that is likely to be representative of the population. However, there will be some sampling error with probability samples. In terms of measurements like means and standard deviations, a sample can never exactly represent the population from which it was drawn. By calculating the standard error, it is feasible to assess the sample's resemblance to the population and derive the relevant conclusions. A huge standard error infers that the mean of the sample population is not equitably spread around the population mean, suggesting that this sample may not be successfully representing the population. A negligible standard error shows that the mean of the sample is clustered closely around the population mean, recommending that the sample effectively represents the population. By increasing sample size, the standard error can be decreased. Using a huge, random sample is the most effective method for reducing sampling bias.

What distinguishes standard deviation from standard error?

The standard deviation is a marker of to what extent a collection of observations deviates from its mean or predicted values, whereas the standard error is basically an estimation of how widely the collection's average deviates from the genuine population average. The standard deviation is invariably greater compared to the standard error. Let's compare the distinctions between them. The most significant distinctions are: Standard error reflects variability across numerous samples of a population, whereas standard deviation depicts variability within a specific subset of that population. The standard error is a statistical descriptive metric that can only be approximated, whereas the standard deviation is an inferential statistic that can be derived from sample data. The formula for standard error is the standard deviation divided by the square root of the sample size. The formula for standard deviation is the square root of the sample variance.

3.5.4 Coefficient of variation

The coefficient of variation (CV) is a statistical metric of relative variability that represents the amount of standard deviation relative to the population mean. It is a common, non-united measurement that tries to determine the variability of dissimilar groups and attributes. It's also referred to as the "relative standard deviation" (RSD). It comes in particularly handy when evaluating the level of variance between two datasets, since the allowable limit inside the variance grows as the standard deviation number rises and the information becomes less exact. Utilizing a simple ratio, the coefficient of variation is calculated. The variance can be estimated by dividing the standard deviation by the sample mean.

$$CV = \frac{\text{Standard deviation}}{\text{Mean}}$$

For instance, the delivery time of a courier service is measured in minutes. The average delivery time is 30 min, with a standard variation of 6 min, and hence, 0.20 is the coefficient of variation for the parcel delivery case. This value represents the magnitude of the standard deviation relative to the mean. Analysts typically represent the coefficient of variation as a percentage, so for this situation of parcel delivery, the standard deviation is 20% of the mean size. If the value is one or 100%, then the standard deviation and the mean are equal. Less than one implies that the standard deviation is less than the mean, while greater than one indicates that it is greater than the mean. In general, bigger values correspond to a higher degree of relative variability.

3.5.4.1 Comprehending the source of variability for analysis

Every statistical study presumes that whatever is seen in the samples could be extrapolated to the entire populace. We might be fairly sure if our findings are generally applicable to a community under investigation if we recognize the samples. We can cope with several causes of variation, including biological as well as technological ones, which, if disregarded, can have a significant effect on the statistical analysis as well as the results. Scientists doing omics studies are familiar with a variety of origins of unpredictability. Details can comprise information regarding dates and manner of data gathering or processing, the procedure followed, the scientist doing preliminary studies, computational tools, and a variety of other discovery details. We can adjust for such possible disparities in the data using both quantitative and technological methods, but we'll require such evidence first. We could set up the study so that all the data is collected in the same way by the same scientist, or we could use controls to make up for these differences in the statistical analyses. Determinants of diversity in the research group are also crucial. These could be unfamiliar to individuals conducting omics research, yet they certainly could have an impact on overall results. When making a prediction based on different factors, like the effect of the drug, it is important that the samples that got the drug and the ones that didn't get it are as similar as possible. Whenever the distinguishing factor is an inherent contradiction across data, such as when comparing the expression of genes in people with the presence or absence of an illness, it gets increasingly problematic. It's hard to locate infected and uninfected people that are otherwise identical, so any changes in expression could be attributable to anything besides the illness.

3.6 Different types of distributions and their significance

A "distribution" is basically a set of data or scores associated with a variable. Typically, these values are sorted from lowest to highest and can then be shown graphically. The most popular distribution is the Gaussian distribution, commonly known as the "normal" distribution. This distribution provides a standardized mathematical formula that may be used to calculate the probability for each sample observation. The probability density function, which indicates the grouping or density of the data, is represented by this distribution. The likelihood that an observation has a value equal to or less than a specified value is also calculable. A cumulative density function offers an overview of various data relationships. In practice, we might think of a distribution as a statistic that shows the relationship between observations in a data point. For example, we may be interested in people's height, with individual heights representing domain observations and heights 140−180 cm being the sample space. The distribution is a mathematical function that depicts the relationship between different height readings. Many data points are congruent with well-known and acknowledged mathematical functions, such as the Gaussian distribution. Modifying the function's parameters, such as the mean and the standard deviation in the case of a Gaussian distribution, enables the function to be tailored to the given data. When a distribution function is known, it may be utilized as a shortcut for defining and computing related quantities, such as observation probabilities and domain connections (Islam and Al-Shiha, 2018).

3.6.1 Probability distributions

The probability distributions are statistical expressions that represent all of the potential readings and probabilities for a random vector inside a particular region. This region would be limited by the lowest and greatest input states, and wherever that potential quantity would be displayed just on the probability density function, it would be determined by a variety of parameters. The average, standard deviation, skewness, and kurtosis of distributions are among such parameters. The normal distribution, also known as the "bell-shaped curve," represents among the most frequent probability distributions; however, there are many more. Usually, this method of obtaining information about a phenomenon determines its probabilistic model. The probability density is the name for such a procedure. In the data design phase, there are two types of probability distributions that are used for different reasons is the discrete probability function, and the second is the continuous probability function (Khakshooy and Chiappelli, 2018).

Some examples of these classes include the normal distribution, chi-square test distribution, binomial distribution, and Poisson distribution. Distinct probability distributions describe distinct information generating methods as well as fulfilling various purposes. The binomial distribution calculates the estimated likelihood of the occurrence of events multiple times across a set of experiments, knowing the overall likelihood of an incident within every attempt. Because just 1 or 0 is a legitimate answer, a binomial distribution is discontinuous rather than being a continuum.

3.6.2 Continuous probability function
3.6.2.1 Normal distribution
The normal distribution is a homogeneous/symmetric probability distribution about the mean that shows data closer to the mean are more common than data farther from

it. A bell curve (Fig. 3.7) appears on a diagram representing a normal distribution. A random variable with a normal sampling distribution (also called a Gaussian distribution) is called a normal deviation. Although all normal distributions are symmetric, not all symmetric distributions are normal. The mean and standard deviation fully characterize the normal distribution, showing that the distribution is unbiased but has kurtosis. The mean (average) equals zero and the standard deviation equals one, which define a normal distribution with zero skewness and 3 kurtoses (Fig. 3.7).

In a normal distribution, approximately 68% of the collected data falls within ± 1 standard deviation of the mean, 95% of the data falls within ± 2 standard deviations of the mean, and 99.7% of the data falls within ± 3 standard deviations of the mean. As a clear contrast to binomial distribution, the normal distribution is continuous in nature, meaning that all possible values are displayed (as contrasted to 0 and 1 with no intermediate value).

When, "$\mu = 0$" and "$\sigma = 1$", it is called as Probability density function denoted by "$f(x)$", "μ" refers to mean (median and mode also) of the distribution and "σ" refers to standard deviation.

$$f(x) = \frac{1}{\sigma\sqrt{2\pi}}\, e^{-\frac{1}{2}\left(\frac{x-\mu}{\sigma}\right)^2}$$

Normal distribution integrals are the Cumulative density function of conventional normal distributions, as depicted in equation below. This is typically represented with a capital Greek letter "Φ".

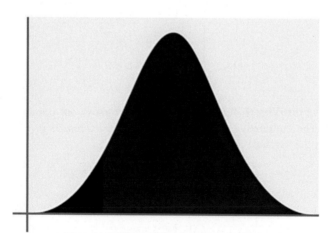

BELL CURVE REPRESENTING STANDARD NORMAL DISRTIBUTION

FIG. 3.7

A bell curve illustrating the conventional normal distribution.

$$\Phi(x) = \frac{1}{\sqrt{2\pi}} \int_{-\infty}^{x} e^{\frac{-t^2}{2}} dt$$

3.6.2.2 Continuous uniform distribution

A continuous uniform distribution is a symmetric likelihood distribution, commonly known as a "rectangular distribution" as it forms a rectangle on a graph. This type of continuous distribution is a method for describing an experiment whose outcomes are random and fall within a given range. Minimum and maximum values are specified via the parameters "a" and "b", which set the limits. The interval range for this distribution can be closed, as in [a, b], or open, as in (a, b). Thus, the distribution is commonly represented as U (a, b), where "U" stands for "uniform distribution." The gap between the interval limits determines the length of the interval. All intervals of the same length are possible on the distribution's support. It is the probability distribution with the highest entropy for a random variable "x" that is not limited in any way other than being part of the distribution.

In relation to a continuous uniform distribution, the probability density function is represented by:

$$f(x) = f(x) = \begin{cases} \frac{1}{b-a}, For\ a \le x \le b, \\ 0, For\ x < a\ or\ x > b \end{cases}$$

The value of "$f(x)$" at the two limits a and b are often insignificant since they do not affect the integrals of "$f(x)dx$" across any interval. Occasionally they are selected as zero, and sometimes they are selected as "$\frac{1}{b-a}$". In the situation of estimating using the approach of maximum likelihood, the latte is suitable.

$$F(x) = \begin{cases} 0, For\ x < a \\ \frac{x-a}{b-a}, For\ a \le x \le b \\ 1, For\ x > b \end{cases}$$

Cumulative distribution function is represented above for continuous uniform distribution. The cumulative distribution function is expressed in the notation of the mean and the variance as:

$$f(x) = \begin{cases} 0, For\ x - \mu < -\sigma\sqrt{3} \\ \frac{1}{2}\left(\frac{x-\mu}{\sigma\sqrt{3}} + 1\right), For\ -\sigma\sqrt{3} \le x - \mu < \sigma\sqrt{3} \\ 1, For\ x - \mu \ge \sigma\sqrt{3} \end{cases}$$

3.6.2.3 Log-normal distribution

A probability distribution with a normally distributed logarithm is called a lognormal distribution, or Galton distribution. A random variable is lognormally oriented if its logarithm follows the same pattern as the normal distribution. This type of distribution often has skewed distributions with minimal mean values, massive variance, and all positive values. Log(x) only works when x is positive, so values have to be positive. The mean "μ" and the standard deviation "σ" tell us what the probability density function for Log-normal distribution is.

A random variable "X" that is positive is said to have a log-normal distribution if its natural logarithm has a normal distribution with the mean and the variance.

$$\ln(X) \sim \mathcal{N}(\mu, \sigma^2)$$

If "φ" represent the probability density function for the distribution $N(0.1)$, then it follows that:

$$f_X(x) = \varphi\left(\frac{\ln x - \mu}{\sigma}\right)\frac{1}{\sigma x}$$

And if "Φ" represent the cumulative distribution function for the distribution $N(0.1)$, then it follows the equation:

$$F_X(x) = \Phi\left(\frac{(\ln x) - \mu}{\sigma}\right)$$

3.6.2.4 Exponential distribution

The time between events in a Poisson point process, where events occur repeatedly, autonomously, and at a steady average rate, has an exponential probability distribution, and this type of distribution is known as an exponential distribution. This is a special example of the gamma distribution. It's the continuous version of the geometric distribution, and the most important thing about it is that it doesn't remember anything. It is used for more than just studying Poisson point processes.

For an Exponential distribution, Probability distribution function can be defined as:

$$f(x; \lambda) = \begin{cases} \lambda e^{-\lambda x}, x \geq 0 \\ 0, x < 0 \end{cases}$$

Here, $\lambda > 0$ is the exponential distribution variable, which is also termed as the rate parameter. The valid range for this type of distribution is between [0 and infinity). For a random variable "X" having an exponential distribution, then we will write $X \sim \text{Exp}(\lambda)$ and the exponential distribution can be divided into infinite number of parts.

Cumulative distribution function can be defined as below for an exponential function.

$$F(x; \lambda) = \begin{cases} 1 - e^{-\lambda x}, x \geq 0 \\ 0, x < 0 \end{cases}$$

3.6.3 Discrete probability function

A discrete distribution reflects the likelihood of occurrence for each discrete random variable value. For example, let's take the example of rotten eggs in a tray. The number of total rotten eggs out of a total number of eggs will be the discrete probability function. In a discrete probability distribution, each possible value of a random variable is associated with a probability that is positive and greater than zero. Some of the significant probability distribution functions are discussed below.

3.6.3.1 Binomial distribution

A binomial distribution could be conceived of as the likelihood of a successful or unsuccessful output in a multi-step investigation or assessment. The binomial distribution is a type of probability distribution that takes into account two different possible outcomes. It's a popular approach to evaluating probability distributions that is utilized in a lot of statistical methods.

In instance, we use X ~ B (n, p) if the random variable "X" approaches the binomial distribution for parameters $n \in N$ and $p \in [0,1]$. The probability mass function tells us how likely it is that we will get exactly "k" successes out of "n" separate Bernoulli trials:

$$f(k,n,p) = \Pr(k;n,p) = \Pr(X=k) = \binom{n}{k} p^k (1-p)^{n-k}$$

for $k = 0, 1, 2, \ldots, n$, where $\binom{n}{k} = \frac{n!}{k!(n-k)!}$ is the binomial coefficient, that's why this distribution is called the binomial distribution. This binomial distribution formula can be explained like this: "k" successes happen with a chance of "p^k", and "$n-k$" failures happen with a chance of "$(1-p)^{n-k}$". The "k" successes, on the other hand, can happen anywhere in the "n" trials, and there are "$\binom{n}{k}$" different ways to spread them out over the "n" trials.

One possible way to express the cumulative distribution function is as follows:

$$F(k;n,p) = \Pr(X \leq k) = \sum_{i=0}^{\lfloor k \rfloor} \binom{n}{i} p^i (1-p)^{n-i}$$

Where, "$\lfloor k \rfloor$" is the "floor" under "k", i.e., the highest integer less than or equal to "k".

3.6.3.2 Bernoulli's distribution

The Bernoulli distribution is a discontinuous likelihood distribution. With probability "p" the Bernoulli distributed random variable has the value 1, but with probability "$q = 1-p$", it has a value of zero. In a less formal context, it can be regarded as a statistical model for the assortment of probable consequences of any individual testing that poses a true-or-false questions. These questions result in boolean-valued responses, where a value is considered "success" or "yes" with a likelihood of "p" and "failure" or "no" with a likelihood of "q". It can be used to epitomize a coin toss, with 1 and 0 signifying "heads" and "tails" and "p" representing the probability that the coin will drop on "heads" or vice versa. If a coin wasn't fair, it would have a probability of "$p \neq 1/2$".

The binomial distribution has several subtypes, including the Bernoulli distribution that only applies when only one trial is done (so for this type of binomial distribution, "n" would be 1). Additionally, this is a specific example of the two-point distribution, in which the potential consequences do not have to be either 0 or 1.

If "X" is a random variable and it follows this distribution, then the following holds true:

$$\Pr (X = 1) = p = 1 - \Pr(X = 0) = 1 - q$$

This distribution's probability mass function "f" across all potential outcomes k is:

$$f(k;p) = \begin{cases} p, k = 1 \\ q = 1 - p, k = 1 \end{cases}$$

3.6.3.3 Poisson distribution

The Poisson distribution is an example of a discontinuous likelihood distribution that is used in probability and statistics. If the rate of events is known to be constant, it illustrates how likely it is that a certain number of actions or events will arise in a definite period of time or place, regardless of how long has passed since the last occurrence. How many events occur in a given distance, area, or volume can be determined using the Poisson distribution. A poisson distribution possesses positive skewness, and whenever the average of a range is large, it resembles a normal or symmetric distribution. A Poisson distribution's form varies; for instance, a lower average Poisson distribution shows severe skewness, featuring 0 as its mode. Having a tail spreading toward the right, every value is forced up against 0.

If "X" a discrete random variable and has a Poisson distribution with the limitation "$\lambda > 0$", then the probability mass function for "X" is given by the following equation:

$$f(k; \lambda) = \Pr(X = k) = \frac{\lambda^k e^{-\lambda}}{k!}$$

Where, "k" is the number of events (k = 0,1,2 ... n), "e" is the Euler's number having a constant value of 2.71828 ...

3.6.4 **Normal distribution and normal curve**

A normal distribution is a type of symmetrical and balanced likelihood distribution around the mean that illustrates that data close to the mean occur more frequently than data far from the mean. On a graph with a Gaussian distribution or a normal distribution, you will observe a bell curve (discussed earlier in the section on probability distributions).

Although all normal distributions are symmetrical, not all symmetrical distributions are normal. The normal distribution's mean and standard deviation clearly describe it, demonstrating that it is not biased but rather exhibits kurtosis. As a result, the distribution is symmetrical and depicted as a bell curve. As long as the standard deviation is equal to one unity with a skew of zero and kurtosis of 3, then the distribution is considered normal.

3.6.5 **Normal curve**

With a high number of observations and a narrow class interval, a symmetrical frequency curve is produced by a histogram with the same frequency distribution of heights. This is known as the normal curve. The frequency distribution is balanced around a single peak, the mean, median, and mode will coincide in the case of a normal curve. It rises progressively from the lowest frequencies at the classification's extremes to the highest frequencies at the classification's apex. This graph is a test of SD to determine whether or not it was computed from a large, random sample representative of the universe. The form of a normal distribution or curve is highly practical and facilitates statistical analysis. It indicates the probability of occurrence by chance or the frequency with which an observation, as measured by the mean and standard deviation, can occur ordinarily in a population. By "normal," we mean normal. A peculiar finding may have a negative prognosis and also be pathological. The phrases "normal" and "abnormal" are not used in clinical situations.

3.6.6 **Asymmetrical distribution**

Distributions of various forms can be observed in nature. We have previously defined "normal distribution." It lays the groundwork for how mean and standard deviation can be used in real life to measure central tendency and variation.

Certain distributions have asymmetries or skews. Based on whether the long tail of the curve is to the left or right of the highest frequencies, they may be skewed to the left or right. Some of them are bimodal, with two peaks, as in the age-based distribution. In such situations, it is likely that two distinct groups of people are mingled together in the populace. Therefore, the sample under research is diverse.

3.7 **Sampling**

It is not practical to enroll every member or sampling unit of the population in a scientific investigation, to examine all of the world's millions of people to determine the prevalence of cancer, or to examine the efficacy of a medicine on every patient with a certain ailment.

In addition, a large number of investigators are necessary to execute this enormous effort, which may reduce the accuracy of a population-wide survey. Their consistency and accuracy may vary, and their acquisition will be expensive, time-consuming, and difficult. Due to these obstacles, we prefer to choose an appropriate sampling strategy. In medical research, sample data are taken from a suitably large and representative population, selected using a normal sampling method. Before drawing a sample, the population must be well-defined. For example, the word "population" can mean a certain group of people, like factory workers, family members, etc., about whom information is needed.

A parameter is a computed value derived from a specific population, such as the mean, the standard deviation, or the standard error of the mean. The value is constant since it applies to the entire population. Statistics, such as mean, standard deviation, and percentage, are values computed from a sample.

There are two primary aims of sampling:

- The evaluation of parameters of the population (mean, proportion, etc.) based on sample statistics.
- To test the population-based hypothesis concerning the sample or samples.

A "sample" is any part of a population that is representative of the whole. Even if a significant number of data samples are gathered from the same group of people, it is still possible that not every single individual is included in the results. The content of samples might vary in terms of size, quality, and sampling procedure; thus, their statistics can also vary. Inferences obtained from a sample pertain solely to the designated population from which the sample or samples were collected. Such findings are applied to the entire population, but generalizations are only true if the samples are sufficiently big and unbiased, i.e., representative of the population from which they are collected. A statistically representative sample will have parameters that are almost identical to those of the total population. There will still be a discrepancy or error of chance, which may be determined from the sample that is representative. The disparities can be diminished but not abolished. A representative sample possesses two primary qualities.

- Precision which means the sample size
- unbiased nature

These two characteristics enable us to accomplish the aforementioned sample objectives. Then, we may determine if observed discrepancies in sample values are attributable to chance or other causes when compared to population characteristics or another sample's statistics.

3.8 Probability

Probability is the measure of the average relative frequency or likelihood of an event's occurrence. The primary objective of choosing a fair sample group is to determine the likelihood (relative frequency) of the occurrence of single or group observations in the normal distribution of any biological variable. We would also like to evaluate the likelihood of sample data points (means and proportions) happening by random chance in order to facilitate the comparison of sample findings to population outcomes. It is therefore possible to rule out random variation and derive conclusions about the population. To do all of this, it is obligatory to have a secure grasp of the probabilities of biological occurrences and events in the population group.

Probability is typically denoted with the sign "p". It varies between zero and one. When "p" equals to zero, there is no possibility that an event will occur, or its occurrence is unlikely. If "p" equals to one, the probability of an event occurring is 100 percent, i.e., it is unavoidable, such as the death of every living creature.

If the probability of something happening in a population is "p" and the probability of it not occurring is given by "q", then

$$q = 1 - p \text{ or } p + q = 1$$

The probability "p" or likelihood of a positive event is calculated using the following formula:

$$p = \frac{\text{Number of events occurring}}{\text{Total number of trials}}$$

It is crucial to have a thorough understanding of probability, as it serves as the foundation for all assessments of significance. It is often approximated using the five laws of probability, the normal curve, and the tables.

3.8.1 Laws of probability
3.8.1.1 Addition law of probability
The concept of mutually exclusive occurrences refers to situations in which the probability of one event eliminates the possibility of another event or events taking place at the same time. When flipping a coin, receiving head eliminates the possibility of obtaining tail. An occurrence will arise in one of the many possible ways. Thus, occurrences that are mutually exclusive adhere to the addition law of probability. If the count of mutually exclusive occurrences is "n" and the individual probability is "p^1", then the overall probability is as follows:

$$P = p^1 + p^2, \ldots\ldots + p^n = 1$$

Whenever addition law is used, the term "or" is present, e.g., birth of Rh-negative or Rh-positive baby, a medicine will cure or alleviate or have no impact on an ailment.

3.8.1.2 Multiplication law of probability
This law is used when two or more events happen simultaneously, but they must not be related, meaning the events should be independent of one another. When two dice

throws result in an event like 3 and 6 or 6 and 3, the word "and" is used between the occurrences.

If a dice is tossed twice in a row, what is the likelihood of having 3 and 6 or 6 and 3, with the likelihood of obtaining 3 in the first toss being 1/6 and 6 in the second being 1/6?

The likelihood of obtaining 3 in the first toss is 1/6 and 6 in the second will be 1/6. So, in first scenario the likelihood of obtaining 3 in the first toss and 6 in the second is $1/6 \times 1/6 = 1/36$.

The likelihood of getting 6 on the first toss and 3 on the second would be $1/6 \times 1/6 = 1/36$ in the second scenario. It demonstrates that order is irrelevant.

The likelihood of obtaining either in two tosses, if the sequences of 3 and 6 or 6 and 3 are ignored, is $1/36 + 1/36 = 2/36 = 1/18$.

In certain situations, it is necessary to use both the addition and multiplication principles, and the conjunctions "and" and "or" must be employed. For instance, the Rh factor and birth sex are separate occurrences that can happen to any infant.

3.8.1.3 Binomial law of probability distribution
Binomial law of probability distribution is established by the terms of the expansion of the binomial expression $(p + q)^n$, where "n" represents the sample size or the number of events such as births, tosses, or randomly selected individuals for whom the probability is to be calculated, "p" represents the probability of 'success or true or 1,' "q" represents the probability of "failure or false or 0" and $(p+q)$ is equal to unity.

For instance, when n = 2, the expansion terms of $(p + q)^2$ are p^2, 2pq, and q^2. The values of "p" and "q" are derived from the percentage of population, i.e., the probability of having a baby boy "p" or a baby girl "q" when just one child is born is determined by watching a huge number of births in the population. It might be 51% for boys (p) and 49% for girls (q). Substitute p = 0.51 and q = (1 − 0.51 = 0.49) to determine the probability.

3.8.1.4 Probability (chances) from shape of normal distribution or normal curve
If the heights of the students in a class follow a normal distribution and the total number of students in the class, let's say 200, is considered to be unity, then we know that 50 percent of the students in the class are taller than the mean and 50 percent of the students are shorter than the mean. The range, mean ±1 SD, includes 68% of children, while the mean ±2 SD covers 95%. Therefore, the likelihood of having a height above the mean with +2 standard deviations is 2.5%, and the probability of having a height below the mean with −2 standard deviations is also 2.5%.

It is possible to estimate the likelihood of any observation or number of instances lying above or below the mean at any distance from the mean. Likewise, the whole area beneath the normal curve is assumed to be unity. The proportionate area under any piece of the curve for a normally distributed variable indicates the relative frequency or likelihood of observations between any two places on the horizontal scale.

Similarly, the sampling distribution determines the probability of sample values or results, such as means and proportions, varying by chance from those of the other samples or the population. Theoretically, if a sample is fairly representative of a population, its mean should be the same as that of another representative sample or the population mean. However, this is not what happens in practice. 95% of sample means fall within the range of a mean ±1.96 standard error, according to the sampling distribution. 5% probability of the data being higher or lower than this range (0.0.5 out of 1).

3.8.1.5 Probability of calculated values from tables

Using the corresponding tables, the probability of computed values occurring by chance for "t" and "χ^2" (Chi-square) is obtained. The probabilities or odds of dying or thriving at any age are derived from life tables based on the mortality experience of a large sample of the population that is inclusive of both men and women of all ages. Additionally, modified life table approaches are utilized to determine the probability of survivability at any given time after a therapeutic intervention or procedure.

3.9 Comparing the means of two or more data variables or groups

By comparing means, the t-test compares the mean of a variable in one group to the mean of the same variable in one, two, or more other sample groups. The null hypothesis is set to zero for the population difference between the two groups. We test this null hypothesis using population sample data. We can perform a one-tailed test (lower than or higher than) or a two-tailed test. For instance, we use one-tailed tests to see if the available data show that the sample mean difference across groups is lower than (or higher than) zero. In many statistical settings, the means of two groups or samples must be compared. The method used to compare means depends on the type of data and how the data is organized (Pereira and Leslie, 2009). These are the four most common ways to compare means for data that is thought to be normally distributed:

3.9.1 Independent samples t-test

The most commonly used type of t-test is the independent samples t-test, which is also called the unpaired samples t-test. Examining the means of two independent sets of sample data can be useful. If two samples are drawn from the same demographic group for the independent samples t-test, their means may be equivalent. When samples are collected from two diverse populations, the sample mean may differ. In this instance, it is employed to form inferences on the means of two populations and to determine their relationship. For instance, a t-test makes a comparison between the average test results of men and women in response to the question,

"Could these differences have happened by chance?" This independent t-test is applicable in two situations: first, when neither the mean nor the standard deviation of the sample population are known. Second, when two samples are distinct and unrelated to one another.

There are few assumptions in independent samples t-test:

- Independence assumption: two independent and categorical classes are required to express an independent variable. In above example, male and females are two well defined categories and hence, constitutes the independent variables.
- Normality assumption: the dependent variable should have a normal distribution. Additionally, the dependent variable must be evaluated on a continuous measure. In above example, average test results are defined as dependent variables.
- Uniformity of Variance Assumption: the variances should be equivalent for the dependent variables.

3.9.2 One sample t-test

It is a statistical test designed to evaluate if the mean calculated from sample data obtained from a single group deviates from a value provided by the researcher. This is an extrinsic value determined for scientific purposes and not obtained from the sample data itself. Typically, this selected value is a previously determined demographic mean, a standard value of concern, or a mean suggested from prior research. The one-sample t-test is like other hypothetical tests in that it checks to see if there is enough information to reject the null hypothesis "H-0" in favor of the alternative hypothesis "H-1." "The population mean is equal to the given mean value," is the null hypothesis for a one-sample t-test, while "the population mean is different from the given mean value," is the alternative hypothesis.

The one sample t-test differs from the vast majority of statistical hypothesis tests in that it neither evaluates the association between variables nor compares two distinct groups. It is a direct measure of comparison between data collected on an individual variable from a specific solitary population and a researcher-determined value.

It is possible to do a one-tailed or two-tailed t-test using the one sample t-test, depending on whether you are looking for a discrepancy from the mean in one direction or in the both directions. An assumption-driven test, the One Sample t-test requires the following:

- Sample data should be independent.
- Sample data is obtained at random as in case of simple random sampling.
- The sample data should have a normal distribution.

3.9.3 Paired samples t-test

The paired sample t-test, also known as the dependent sample t-test, is a statistical metric tool for estimating if the mean difference between two sets of observational data is equal to zero. In this test, each object or entity is measured twice, resulting in two sets of observations. The paired sample t-test is widely used in case-control studies and designs that require repeated measurements. Suppose you wish to determine the efficacy of a program for training students or groups of patients. Using a paired sample t-test, you may examine the differences between the performance of a sample of students or the efficacy of therapy before and after the completion of the process and then look at the significant differences.

The paired sample t-test employs two contradictory research hypotheses, the null hypothesis and the alternative hypotheses, just like most statistical procedures. The null hypothesis states that there is no difference between the mean values of the two paired datasets. According to this perspective, all visible distinctions are the result of random variation. It is possible that the real mean difference between the two samples is not equal to zero, as the alternative hypothesis asserts. The structure of the alternative hypothesis can vary based on the predicted outcome. A two-tailed hypothesis is utilized when the direction of the difference is irrelevant. Alternately, upper- or lower-tailed hypotheses might be used to enhance the validity of the test. Below are the basic definitions of the hypotheses for the paired sample t-test.

- According to the null hypothesis (H^0), the actual mean difference (μ^d) equals 0.
- The alternative hypothesis with two tails (H^1) indicates that (μ^d) is not equal to 0.
- The alternative hypothesis with an upper tail (H^1) indicates that (μ^d) exceeds 0.
- The alternative hypothesis with a lower tail (H^1) indicates that (μ^d) is less than 0.

3.9.4 ANOVA

Analysis of variance (ANOVA) is an ensemble of statistical theories and their corresponding estimation procedures (such as "variation across or between groups") that are used to determine the differences between means. ANOVA is governed by the law of total variance, which partitions the observed variance in a variable into components associated with various sources of variation. In its most fundamental form, ANOVA provides a statistical test to evaluate if two or more population means are equivalent, hence expanding the scope of the t-test beyond two means. In other words, when two or more populations have different mean values, ANOVA can be used to determine how significant the difference is.

The assumptions behind the ANOVA test are much the same as underlying for any parametric test.

- ANOVA can only be performed if the subjects in every sample are unrelated. This indicates that subjects in the first group can't be present in the second group (i.e., independent samples/between-groups).
- The sample sizes of the distinct groups must be equivalent.
- An ANOVA can be performed only if the dependent variable has a normal distribution, in which the middle values are most prevalent and the extreme values are rarest.

- Population variances must be homoscedastic means they should be identical. Homogeneity of variance indicates that the deviation of values (as measured by range or standard deviation) is comparable across populations.

There are different types of ANOVA tests. "One-Way" and "Two-Way" are the most popular types. The distinction between these two categories is controlled by the number of independent variables in an ANOVA test. A one-way ANOVA consists of one categorical independent variable (also called a factor) and one normally distributed continuous dependent variable, for example, testing the effect of a treatment intervention on the likelihood of anxiety in a clinical sample. But a two-way ANOVA consists of two or more categorical independent variables and a single normally distributed continuous dependent variable, for example, employment status is an example of a factorial two-way ANOVA.

3.9.5 **The Chi-square tests**

The Chi-square test is a statistical way to compare two sets of data that doesn't depend on any assumptions or variable distribution. Despite its uniqueness, the Chi-square test has a well-defined distribution that makes it a significant tool for scientific research. It is utilized most frequently when data consist of frequencies, such as the number of responses in two or more population groups. It has three common but crucial uses in biomedical statistics:

- Proportion
- Interrelation
- Integrity of fit

The Chi-square (χ^2) test is a valuable tool for comparing experimentally obtained findings to those theoretically predicted by a given hypothesis. Consequently, the actual difference between the observed and the expected frequencies is captured by the Chi-square statistic. To understand the discrepancy between theory and reality in sampling research, a measurement like this is essential to use. According to the definition, it is as follows:

$$\chi^2 = \sum \frac{(O - E)}{E}$$

where, "O" represents observed frequencies and "E" represents expected frequencies.

Calculating the standard error of difference between two proportions would reveal the relevance of large binomial samples with a size greater than 30. Chi-square test is another extremely helpful test that may be used to determine significance in the same type of data, with two additional benefits.

- To compare the results of two binomial samples, even if the samples are small, fewer than 30.

- To evaluate and compare the frequencies of two multinomial samples such as number of diabetic as well as non-diabetic in different weight groups of individuals.

The Chi-Square test has numerous uses when other parametric tests cannot be utilized. Some applications of Chi-square test are summarized with examples below:

3.9.6 Test of independence

This test aids in identifying the relationship between two or more features. Suppose N observations have been classified according to two features. By implementing this test on the observations (data) provided, we attempt to determine if the features are associated or independent. This relationship could be good, negative, or nonexistent. For instance, we can determine whether there is a correlation between punctuality in class and the percentage of students who pass, and whether paracetamol is successful in reducing fever. The null hypothesis that there is no association is tested to see if the variables under investigation are linked. To put it another way, the two characteristics are distinct from one another.

Once the value of chi square has been calculated, it is compared to the critical threshold value for the given degree of freedom at a certain significance level. As long as the predicted chi-square value is below the critical or threshold table value, we can accept the null hypothesis and say that two characteristics are not connected. If the calculated number is higher than the value in the critical or threshold table, the experiments have shown that the hypothesis is false, so the hypothesis is rejected.

3.9.7 Test of goodness of fit

This is the most valuable application of the Chi Square test. This method is utilized mostly for checking the goodness of fit. It seeks to establish whether an observed frequency distribution is distinct from an estimated frequency distribution. When a normal or other sort of ideal frequency curve is fitted to the data, it is important to determine how well this curve corresponds to the observed data.

The following steps are taken to attain the aforementioned objective: In relation to the inquiry, a null and alternative hypothesis are formulated, and a significance level for rejecting the null hypothesis is chosen. A random observational sample is drawn from a suitable statistical sample. Probability is used to calculate theoretical or expected frequencies based on actual observations. Typically, this includes assuming that a particular probability distribution applies to the statistical sample under study. After that, the observed frequencies are compared to the anticipated or theoretical frequencies. If the estimated value of the chi-square is less than the table value at a given level of significance (typically 5%) and for specific degrees of freedom, the fit is regarded as satisfactory. Thus, the discrepancy between the observed and predicted frequencies can be attributed to sampling variations. Alternatively, if the estimated value of the chi-square is greater than the table value, the fit is regarded as inadequate, i.e., it cannot be attributed to the fluctuations of simple sampling, but rather to the theory's failure to account for the observed facts.

3.9.8 Correlation and regression

Correlation and regression are the two most involved strategies for investigating the connection between two quantitative factors. Most of the time, correlation is seen as the question of whether or not there is a link between two things ("x" and "y"). The Spearman's correlation coefficient rho and the Pearson's product-moment correlation coefficient are the two most renowned correlation coefficients.

Conversely, regression gauges the value of the dependent variable in light of the observed value of the autonomous/independent variable, considering the typical measurable statistical connection between at least two factors. There is a lot of equivocalness in figuring out the thought.

3.9.9 A look into correlation and regression

For example, when one variable changes, the other also changes. As a result, either directly or indirectly, a shift occurs. It is possible to have uncorrelated variables if one variable does not move in the same direction as another variable. Calculating the degree to which two variables are interconnected is done statistically. It's possible to have both positive and negative correlations. A rise in one variable leads to an equal and opposite increase elsewhere; this is a sign that the two variables are positively linked when they move in the same direction at the same time. As a case study, profit and investment can be cited. The converse is true when two variables move in opposite directions, such as when one rises and the other falls. This scenario is characterized by a negative correlation. Price and demand for a product are excellent examples of this.

When one or more independent variables are altered, Linear Regression can be used to estimate the change in the dependent variable's metric as a result of the change in the average statistically significant association between two or more variables. Linear regression is a powerful and versatile method for predicting the past, present, or future based on past or present occurrences that is used in a number of human activities. An example of this can be found in the following statement: The profits of a company can be calculated using the company's previous results. It is important to note that in a simple linear regression, there are two variables "x" and "y", that are dependent on or influenced by each other. In this context, "y" is referred to as a criterion or dependent variable, whereas "x" is a predictor or independent variable. Below equation provides the regression line for "y" with respect to "x":

$$y = a + bx$$

Here, "a" is a constant and "b" is a regression coefficient.

Correlation and Linear Regression are used when there are two things to measure, like food quantity and the weight; medicine dose and its impact on blood pressure; temperature of the air and its influence on metabolic rate. Another nominal variable, like the name of a species, a study, or a location, keeps the two measurements together.

3.10 Platforms employed for statistical analysis

The majority of the processes employ a Linux system in conjunction with R programming and integrated development environments (IDEs) such as Rstudio, Jupyter Notebook, Vim, etc. For analyzing Next Generation Sequencing (NGS) research, the accessible user interface options are constrained, with Galaxy being by far the only credible choice. DNAnexus as well as artificial intelligence-based RNA-seq (AIR) are two cloud software UIs that are both patented and subscriber based. Massive volumes of genetic data can also be processed using cloud services like Amazon Web Services and Microsoft Azure, however, these are largely controller driven.

3.10.1 Downstream analysis and visualization

Multiple downstream assessments can be done once the study to determine substantially expressed profiles is completed for adding biological context against the described genomes. The two major downstream analysis performed on transcriptome dataset are as follows:

3.11 Gene ontology & pathway analysis

To have a deeper understanding of the biologic systems being investigated in a specific study, one must first comprehend the role and routes by which a gene is expressed in that system. The functionality of genes could be classified using Gene Ontology (GO) categorization methods into 3 groups: cellular components, molecular functions, and biological processes. The GO study, called gene enrichment analysis (GEA) may alternatively be classified into three groups:

3.11.1 Singular enrichment analysis (SEA)

An entry could be a user-defined gene array taken from any high-throughput NGS/microarray study. Depending on the user-defined arbitrary standard, genome inputs are frequently the most important. The genes are classified under the three major areas mentioned before. Fisher's exact test and its variations, such as the EASE score or the Chi-square test, are used in this method to find out if the genotypes are linked to the functional classification. SEA is performed using techniques like DAVID, GoStat, and Bingo.

3.11.2 Gene set enrichment analysis (GSEA)

All genomes within a high-throughput genome study are utilized as inputs for genome set enrichment analysis, guaranteeing that the assessment is independent of any possible bias owing to artificial limits like those employed in SEA. Perhaps genes with minor differentially expressed variations can be evaluated for any further enrichment analysis as a result of this. The ranking of all genome components inside an annotating class is used to determine maximum enrichment scores (MESs). These maximal enrichment scores are evaluated, and the "p-value" is usually found by using Kolmogorov–Smirnov type statistics to compare reported MESs with ones given in a probability sampling or parameterized statistics to measure fold variations across trials. Such an assessment can be done with ErmineJ and FatiScan.

3.11.3 Modular enrichment analysis (MEA)

To facilitate phrase linkages, modular enrichment analysis blends SEA type enrichment analysis alongside connectivity search techniques. With this investigation, Kappa estimates of concordance were employed. Genes that do not appear in numerous nearby words are removed from study considering the nature of a Kappa statistic. ADGO, DAVID, as well as GeneCodis, are a few platforms that can execute this type of research. Such gene ontology categorization methods integrate information from a variety of domains, such as KEGG in pathway assessment, Pfam in the case of protein domains, or TRANSFAC for transcriptional regulation.

3.11.4 Correlation networks

One important insight gained from the genes is whether or not there is any statistically significant association between the genome list.Even though GEA provides a humongous amount of information regarding genes having similar functions, programmes like GeneMania offer information on gene interactions, resulting in a much more thorough view of the interaction patterns. GeneMania predicts co-expression, co-localization, as well as physical forces amongst genomes, for its users. One such resource is the Biological General Repository for Interaction Datasets (Biogrid), which contains biochemical, genomic, as well as protein-protein interaction data based on known test findings and therefore is updated on a frequent basis. STRING is a library of proteins that interact with each other. It is being used to learn more about how proteins made by genes work together. Weighted gene correlation network analysis (WGCNA), an R-tool that analyses expression profiles either using microarrays or RNA-sequencing to build a correlation network among genes in a specific study, is another option.

3.12 Future prospects and conclusion

In bioinformatics, statisticians have made a key contribution to the development of sophisticated modeling and analytic methods that users may employ to derive relevant biological contents from the vast amounts of multi-platform genomic information. Their in-depth knowledge of the research methods, along with their unpredictability and ambiguity, have particularly qualified them to play a key role in this endeavor. Further research is certainly required in a variety of domains, and new progress is conceivable. Integrative research is another important field. This discipline is just getting underway, and the research establishment is in desperate need of innovative approaches for incorporating data across many systems in order to acquire a much more comprehensive understanding of the cellular biology that underpins it. Such techniques must strike a balance between statistical rigor in constructing links, computing effectiveness in scaling up to massive data environments, and the understandability of outcomes so that our colleagues can understand them effectively.

The statistical evaluation of omics studies is a rapidly expanding area in which agreement on the best techniques might be difficult to come by. Validated experimentally for different testing adjustments that are best suited for relevant outcomes would be created as the study proceeds, and so would improved techniques of data incorporation across diverse resources and systems. Nevertheless, combining clinical and genetic information and fully evaluating hypotheses so that the final results are useful to the overall public remains a significant hurdle in biological information processing. Bioscience and health care have come a long way to where big data is now a normal part of research and treatment. Because of this, statistics has a unique chance to make a big difference in the progress of science by giving other researchers tools that need to have important information in them.

References

Dakhale, G.N., Hiware, S.K., Shinde, A.T., Mahatme, M.S., 2012. Basic biostatistics for postgraduate students. Indian J. Pharmacol. 44, 435. https://doi.org/10.4103/0253-7613.99297.

Hazra, A., Gogtay, N., 2016. Biostatistics series module 1: Basics of biostatistics. Indian J. Dermatol. 61, 10. https://doi.org/10.4103/0019-5154.173988.

Islam, M.A., Al-Shiha, A., 2018. Foundations of Biostatistics, pp. 1–459. https://doi.org/10.1007/978-981-10-8627-4/COVER.

Khakshooy, A.M., Chiappelli, F., 2018. Practical biostatistics in translational healthcare. Pract. Biostat. Transl. Healthc. 1–227. https://doi.org/10.1007/978-3-662-57437-9/COVER.

Krzywinski, M., Altman, N., 2014a. Points of significance: analysis of variance and blocking. Nat. Methods 11, 699–700. https://doi.org/10.1038/NMETH.3005.

Krzywinski, M., Altman, N., 2014b. Visualizing samples with box plots. Nat. Methods 11, 119–120. https://doi.org/10.1038/NMETH.2813.

Pereira, S.M.C., Leslie, G., 2009. Hypothesis testing. Aust. Crit. Care 22, 187−191. https://doi.org/10.1016/J.AUCC.2009.08.003.

Royston, P., 1992. Which measures of skewness and kurtosis are best? Stat. Med. 11, 333−343. https://doi.org/10.1002/SIM.4780110306.

Smucker, B., Krzywinski, M., Altman, N., 2018. Optimal experimental design. Nat. Methods 15, 559−560. https://doi.org/10.1038/S41592-018-0083-2.

Algorithms in computational biology

4.1 Sequence alignment

In the discipline of bioinformatics known as "sequence alignment," researchers are working to create computer-based methods for comparing and locating amino acid or DNA base pair sequences that are comparable. The most fundamental aspect of manipulating biological sequences is alignment, which finds use in a variety of contexts, including sequence assembly, sequence annotation, structural and functional predictions for genes and proteins, phylogeny, and evolutionary research. Maybe in 1966, the edit distance between two strings was defined as the smallest number of insertions, deletions, and letter swaps needed to change one string into another. This gave rise to the computer challenge of sequence alignment (Saloom et al., 2022).

4.1.1 Local alignment

Local alignment is possibly the simplest technique to compare two sequences; it is the approach that employs the fewest assumptions regarding how the similarity should be structured: one is tasked with locating any subsequences with a similarity greater than a predetermined threshold. Finding evolutionarily constrained elements, all pairs of genes, other repeats, including transposons, as well as any other commonalities is rather easy using this approach. Local alignments seem to be more useful when two different sequences are thought to have parts that are the same or have similar patterns. Local alignments get rid of any parts that don't match and only align the parts that stay the same between two sequences. This makes sure that the mismatched parts don't show up in the final result.

4.1.2 Global alignment

It is an attempt to align the whole sequences that are similar to some extent and have approximately the same length, for example, orthologous and paralogous sequences. In it, the complete sequence of protein and nucleotides is compared from beginning to end, and vertical bars present between the amino acids represent the identical residues. The classical technique for global alignment is known as the Needleman-Wunsch

All About Bioinformatics. https://doi.org/10.1016/B978-0-443-15250-4.00004-6

algorithm and is dependent on dynamic programming. The algorithm is used to align strings of proteins and nucleotide sequences. The advantages of global alignment are: It is easy to understand, provides complete sequence pairing in output, can examine the small variations between two sequences, and also helps find polymorphism between two sequences. The limitation of global alignment is that in divergent sequences or strings of different length, this method is not capable of giving an exact result as it fails to identify extremely similar local sites among two sequences (Sun et al., 2018).

4.1.3 Gap penalty

4.1.3.1 Gaps and gap penalties

Aligning DNA or protein sequences allows us to estimate their distances, which is useful for inferring molecular function by discovering similarities with known functions. For this reason, a scoring system is required to determine whether an alignment is a "good alignment" of sequences. Therefore, we may learn what kinds of alignments to examine and assess the alignment's significance by using alignment scoring. In the scoring system, there are three outcomes: a match, a mismatch, and a gap (Fig. 4.1). Matches are rewarded with positive values or high scores, while mismatches are penalized with negative values or low scores. This seems pretty easy to do when it comes to DNA, but scoring protein sequences is harder because alignment scoring has to take into account the physiochemical properties of amino acid residues.

While the concepts of match and mismatch are simple to comprehend, the notion of adding gaps is complex. A gap in the sequences indicates the deletion or insertion of amino acid residues or nucleotide bases. When inserting a gap, the following issues could come up: How many gaps are allowed? How long can they last? How do you choose where to put them? Supposedly, it can be attempted to enhance the alignment score by inserting several gaps, but would this be biologically significant? It would seem logical that there must be a problem with this strategy. When we

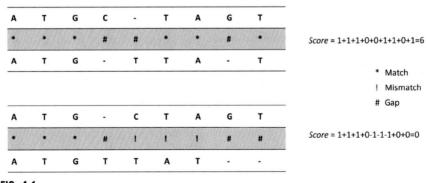

FIG. 4.1

Representation of match, mismatch and gap.

examine dynamically produced sequence alignments, the number of gaps is always restricted. It appears that the algorithms are given instructions to limit the frequency of gaps and their location in the sequence. This directive is referred to as the "gap penalty." Every time the algorithm inserts a gap, it generates a penalty score, which can either lower or raise the alignment's overall score. Each time an alignment is performed, the gap penalty can be modified as a separate parameter. The total number of gaps, their length, and their location in the sequence alignment can all be altered by changing the value of the gap penalties. The match, mismatch, and gap penalties are correlated in the following way: The mismatch and match scores can be used to determine the sequence alignment and its statistical significance if the gap penalty is higher than those scores. If a significant penalty is assigned for both gaps and mismatches, then only matches will be evaluated and checked for significance. While the longest common subsequence may be determined if the gap penalty is zero and the match score is much lower than the mismatch, testing the significance of this subsequence is typically challenging (Fig. 4.2).

There are three types of gap penalty:

Constant or fixed gap penalty: This is the most common and basic sort of gap penalty, where a constant negative score is awarded to the gap irrespective of its length. This kind of gap penalty pushes the algorithm to leave bigger continuous areas and create fewer, larger gaps. For instance, a sequence has three gaps and four matches, and the score assigned to all the gaps is -1 rather than being given individually. In this case, the sequence's overall alignment score will be $4-1 = 3$.

Linear gap penalty: There is an individual score for each gap in this part. The length (L) of each indel in the sequence is considered by the linear gap penalty. If the penalty for each indel residue is A and the gap length is L, the total gap penalty will be the product of the two "AL." So, using the constant gap penalty as an example, the total alignment score of the sequences will be $4-3 = 1$. Because the overall score declines with each new gap, this approach is more favorable for sequences with gaps that are shorter in length. As we increase the gap penalty, the score will immediately increase as well, resulting in alterations to the alignment's significance.

Affine gap penalty: the most often employed gap penalty that penalizes indels. The affine gap penalty combines the variables of both linear and constant gap

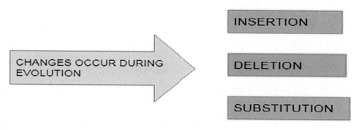

FIG. 4.2

Gap penalty.

penalties, yielding the form X + AL, where L represents the gap length and X and A represent gap opening and gap extension penalties, respectively. Opening penalty is assigned when a gap is first introduced in the alignment of the sequence, if this value is increased, the gaps will be less frequent. The "gap extension penalty" is referred to as the penalty for whenever the gap is extended by one or more residues after gap opening. The gaps will become shorter if the extended penalty value is increased. There will be no penalty for terminal gaps.

The values of A and X are not always obvious since they depend on context. If the goal is to locate matches that are closely connected to one another, it is correct approach to employ a greater gap penalty so as to minimize the number of gap openings. On the other side, when the goal is to find a more distanced match, the gap penalty should be decreased. The size of the gap is also impacted by the relationship that exists between X and A (Fig. 4.3).

4.2 **Pair-wise alignment**

PSA is a technique to align two sequences that searches for the best and most efficient pairwise alignments of a few query sequences using a database similarity search tool. The method has found widespread application in the study of sequences for their functional, evolutionary, and structural properties. When matched sequences reveal a high degree of similarity, the two sequences can be considered members of the same family. Pairwise alignments are used to compare only two sequences at once. They are easy to calculate and are usually used for tasks that don't require a high level of accuracy.

In order to align anything less than an exact alphabetic match, the algorithm must be aware of what it is looking for and how to evaluate the significance of what it finds. In order to do this, "comparison matrices" have been developed, defining a value for each and every potential match scenario—effectively a score of how

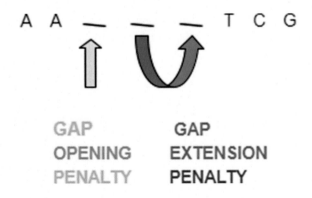

FIG. 4.3

Description of GAP opening penalty and gap extension penalty.

well the computational alignment is performing. The algorithm will look for the best possible score. The total score can only be used for the alignment it creates. It can't be used for anything else.

The aim of pairwise sequence alignment is to find the best pairing of two sequences. K-tuple method, dot-matrix technique, and dynamic programming are the three most used approaches for pairwise alignments. All three approaches have pros and cons, but they all have trouble matching highly repeated sequences with little relevant information, especially when the number of repetitions in the two sequences to be aligned is different.

4.3 Dot-matrix method

A dot matrix approach compares two sequences for potential alignment. In this simplistic approach, the two sequences to be matched are laid out as axes on a grid, and a dot is added at each position where they are an exact match. The top of the matrix lists the first sequence, indicated by the letter X, while the left side lists the second sequence, indicated by the letter Y. Beginning with the first character in Y, one proceeds across the column while remaining in the first row and adding a dot in each column when the character in X is the same. The procedure is repeated once all potential comparisons between X and Y have been performed. The sections of a matching sequence will then be graphically represented by diagonal lines. Any region of similarity may be identified with the use of a diagonal line of dot markers. Dots that aren't on the diagonal but are otherwise isolated signify random matches.

The dot-matrix technique, which generates a group of alignments for given individual sequence sections, is qualitatively and theoretically simple, but evaluating it on a large scale takes a long time. The dot matrix method is computationally available as "Emboss DOTMATCHER," where dot plots can be made as easily as possible. A dot-matrix plot can be used to visually identify sequence properties like insertions, deletions, repetitions, and inverted repeats in the absence of noise. Some implementations change the size or intensity of the dot depending on the degree of similarity between the two characters to accommodate conservative substitutions. The dot plots of closely related sequences will be combined into a single line along the matrix's primary diagonal. Filtering out random matches with a sliding window can make it easier to locate regions that match. It means that more than one nucleotide or amino acid position can be aligned at the same time, and that a dot will only be given if a certain number of matches can be made.

Dot plots have drawbacks as a method of displaying information, including noise, a lack of clarity, unintuitiveness, and difficulties in obtaining information on match positions and summary statistics for the two sequences. There is a lot of unused space in dot-plots since they can only display two sequences, and because the match data is automatically reproduced across the diagonal, noise or empty space takes up a large portion of the plot's real size. By comparing each character using a threshold value and window size, the noise problem may be resolved (the size can be assigned according to the requirement). For a dot to be drawn, a certain number of the matches in the window must be right.

The amount of repetition in a sequence may also be assessed using dot plots. When plotted against themselves, parts of a sequence with a lot in common appear as lines off the main diagonal. When a protein contains several similar structural domains, this phenomenon might occur.

The interpretation of dot matrices is as follows: areas of resemblance will show as diagonal runs of dots; inversions will be indicated by reverse diagonals that are perpendicular to the diagonal; and palindromes will be indicated by reverse diagonals that cross the diagonal (Fig. 4.4).

Dot plots provide several benefits, including the fact that they are quite simple to implement. Its presentation makes it simple to comprehend. It illustrates every combination of aligned pairs that is feasible. It is possible to employ it in conjunction with other different approaches; it finds inverted and direct repeats, insertions, and deletions much easier than the other, more automated approaches do. One drawback shared by most dot matrix computer programmes is that they do not display an actual alignment. There is no score that is returned to show how 'optimal' a certain alignment is (there is no statistical significance that could be examined) (Rice et al., 2000).

4.4 Dynamic programming

Dynamic programming is a technique for segmenting longer sequences into manageable chunks, with each transition between pairs of characters in an alignment accounting for all potential modifications.

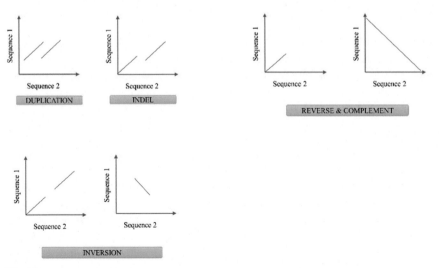

FIG. 4.4

The patterns for interpretation of dot plot results.

In the field of computational biotechnology or bioinformatics, dynamic programming has emerged as the dominant paradigm. Common uses of this information include protein-DNA binding, protein folding, gene recognition, RNA structure prediction, and sequence alignment. The exon/intron structure of eukaryotic genes is determined using DP, which is also applied for assembling DNA sequence information from the segments provided by automated sequencing.

Dynamic programming is based on the idea that solutions to smaller problems can be saved and used later in the larger problem. Comparing this solution with the next solution in the series is normal protocol; the ideal solution will then be compared with the next solution, and so on, till the entire sequence has been exhausted. An algorithm for dynamic programming consists of four components: a recursive formulation of the optimum score; a DP matrix for storing subproblem optimal scores, a method of filling the matrix from the bottom up by tackling the most elementary subproblems first; and a method of tracing back through the matrix to identify the details of the best solution that produced the highest score (Naghibzadeh et al., 2021).

The process of matching nucleotides to protein sequences, which is complicated by the need to take care of frameshift mutations, can be aided by dynamic programming. The method is particularly helpful for sequences with many indels since it may detect frameshifts offset by any number of nucleotides, which makes it challenging to align using more effective heuristic techniques. In practice, the approach requires a system built expressly for dynamic programming or a lot of computing power. The programmes BLAST and EMBOSS offer fundamental tools for producing translated alignments. More broad techniques are offered by the open-source application GeneWise. Given a certain function of scoring, the dynamic programming approach is meant to produce the best alignment; however, selecting a suitable function of scoring is often an empirical rather than a theoretical procedure. Dynamic programming is slow when it has to deal with a lot of sequences or sequences that are too long, even though it can include more than one sequence.

Two classical algorithms of dynamic programming are:

Needleman and Wunsch (1970)—For global alignment and the result contains all residues in the alignment (Needleman and Wunsch, 1970).

Smith and Waterman (1981)—For local alignment and the result contains only certain parts of our sequences (Smite and Waterman, 1981).

4.4.1 Needleman-Wunsch

A bioinformatics technique called the Needleman-Wunsch approach is used for the purpose of aligning protein or nucleotide sequences. When comparing biological sequences, this was one of the first instances of "dynamic programming." The method was developed by Saul B. Needleman and Christian D. Wunsch, and it was first presented in the year 1970. The approach divides a significant challenge, such as the whole of the sequence, into a series of more manageable issues, and then makes use of the solutions found for the more manageable issues to locate an optimal response to the significant challenge. It is also known as the optimum matching algorithm and the global alignment approach. Even today, the Needleman-Wunsch method is frequently utilized, particularly when the quality of the global alignment must be maintained. The technique gives a score to every possible alignment, and its goal is to identify all of the alignments that have the greatest significant value.

The first component of Needleman and Wunsch's ultra-algorithm generates all possible alignments between any pair of sequences, considering their probabilities of being similar, distinct, or containing some insertions along with deletions. When the first phase is complete, a scoring mechanism is implemented to assign a value to each base pair or nucleic acid combination. For example, if 2 Gs are aligned, they receive a score of 1 because they are a perfect match; if C and T are aligned, they receive a value of -1 because they are incompatible; however, if there is no base pair or amino acid available, it is termed a "gap penalty" and receives a score of 0.

Following the completion of all of these steps, the total scores need to be summed in order to select the best alignment possible from among all of the potential alignments that were generated by the procedure. It is necessary to select the alignment with the highest possible score.

Creating a two-dimensional (2D) matrix using the penalty scores for the match, mismatch, and gap is a necessary step in the process. The matrix is solved in three stages: the first stage is called initialization, the second stage is called matrix filling, and the third stage is called traceback (Fig. 4.6).

Seq1: ATTAC
Seq2: AATTC

4.4.1.1 Step 1: Initialization table "T"

The first step in the algorithm is the initialization, in which a scoring matrix is made. The formation of a scoring matrix begins with the placement of sequences, which are placed on the x and y axes of the matrix.

While Seq1 "ATTAC" will be positioned at the "x" coordinate, Seq2 "AATTC" will be positioned at the "y" coordinate. The first column and row of the matrix are initially started from the (0,0) cell, with 0 being the first value in the cell (Fig. 4.6). The gap score is added to the adjacent cell of rows and columns.

4.4.1.2 Step 2: Filling the matrix

The second and arguably most important stage of the process is filling up the matrix, beginning in the top-left corner. It is necessary to know the scores of the cells on the diagonal, left, and right in order to get the maximum score of each cell. Match or mismatch (assumed) scores are added to the diagonal score. In a similar fashion, the gap score is added to the values coming from adjacent cells or boxes (horizontal and vertical). With these three numbers at your disposal, the highest possible score can be obtained; use that to fill the ith and jth slots.

If a value comes from the diagonal direction or cell, it is determined if the residues are identical or not. They are added based on their alignment match or mismatch score. As an outcome, three values come from three independent directions. However, since maximum match alignment is required, only the highest value is placed in the cell (Fig. 4.5)

- Scoring method:
  1. +1 score for every match found.
  2. −1 score for every mismatch.
  3. −1 gap penalty for every insertion and deletion.

4.4.1.3 Step 3: Traceback

After the matrix has been filled in, the final score to be computed is the score of the best possible alignment of the whole sequences. Nevertheless, the best alignment has yet to be determined. This is found by a recursive matrix "traceback." The last and most important step in the sequence alignment process. Starting with the bottom right corner cell, the algorithm determines which of the three highest values was used to fill this cell, and the direction from which that value came is highlighted or saved with a back arrow, before moving to that cell to find the best path or alignment. To fully construct the optimum alignment, the procedure is repeated until the cell (0,0) is reached.

$$
\text{Recursion:} \quad D_{i,j} = \max \begin{cases} D_{i-1,j-1} & + & s(a_i, b_j) \\ D_{i-1,j} & + & s(a_i, -) \\ D_{i,j-1} & + & s(-, b_j) \end{cases} = \max \begin{cases} D_{i-1,j-1} & + & 1 & a_i = b_j \\ D_{i-1,j-1} & + & -1 & a_i \neq b_j \\ D_{i-1,j} & + & -2 & b_j = - \\ D_{i,j-1} & + & -2 & a_i = - \end{cases}
$$

FIG. 4.5

Function for matrix filling.

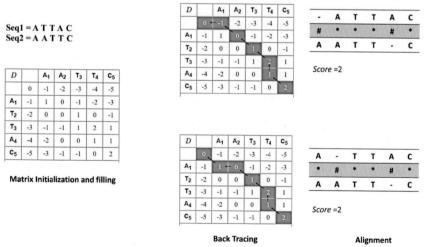

Seq1 = A T T A C
Seq2 = A A T T C

Matrix Initialization and filling

Back Tracing

Alignment

FIG. 4.6

Needleman Wunch algorithm steps.

4.4.2 Smith Waterman algorithm

Temple F. Smith and Michael S. Waterman were the ones who published the strategy for the first time in 1981. The global alignments are produced via the Smith-Waterman method. Protein alignments commonly employ a substitution matrix to provide points for amino acid matches and mismatches, as well as a gap penalty for connecting an amino acid from one sequence to a gap in another. Although in practice a positive match score, a negative mismatch score, and a negative gap penalty are more prevalent, a scoring matrix can be utilized in the alignments of DNA and RNA. However, such results may be explained by altering the procedure. The application of two distinct gap penalties for opening and extending a gap is a frequent expansion of standard linear gap expenditures. The former is typically far higher than the latter; for example, -10 for a gap that is open and -2 for a gap that is extended. As a result, gaps between residues and gaps in alignments are often kept to a minimum, which makes biological sense.

Smith-Waterman's algorithm deals with a segment of the total gene. Rules for filling the scoring matrix in local alignment are the same as global alignment, except that instead of putting values in scoring matrices in real values like -3, -1, -5, $+2$, $+4$, the program puts values only in binary numbers, i.e., 1 or 0. Any negative value in the scoring matrix will be represented by a 0, and any positive value will be represented by a 1.

The sequences were given the following scores: match $= +1$, mismatch $= -1$, and gap $= -1$. All the negative resultant values will be represented as zero, such as in the cells of T1 and G1. However, all the positive values remain the same as calculated. The next step is trace backing, starting from the highest value and tracing backing until the value of zero is not obtained in the first row or column. In the given example, the highest value is "2" in three different cases, the first being C3 and C2, the next being C3 and T3, and the last being C3 and A4. Three different alignments can be seen, and the optimal one can be chosen (Fig. 4.7).

4.4.2.1 Limitations

The exponential growth of genetic data puts the many approaches to DNA sequence alignment that are now in use to the test. In order to fulfill the essential criteria for a method that is both rapid and trustworthy, real-time parallel processing must be accomplished using innovative methods. It has been suggested that optical computing approaches might serve as potential replacements for the electrical implementations that are now in use. One of these methods, known as OptCAM, has been shown to be much more efficient than the Smith-Waterman algorithm.

It was assumed that local alignment shows better alignment when compared to global alignment as they produce patterns that are conserved in whole DNA and amino acid sequences. Researchers say that local alignment can be used in place of the Needleman-Wunsch algorithm for making the different DNA and amino acid sequences, which could consist of a small, matched region or when they are of different lengths, when they are overlapping, or when one sequence is a subsequence of another (Mullan, 2006).

Seq1 = G C T A
Seq2 = T G C

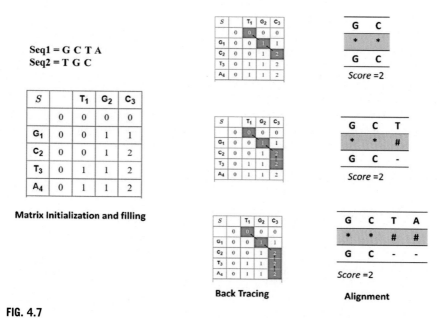

S		T_1	G_2	C_3
	0	0	0	0
G_1	0	0	1	1
C_2	0	0	1	2
T_3	0	1	1	2
A_4	0	1	1	2

Matrix Initialization and filling

Back Tracing   **Alignment**

FIG. 4.7

Smith Waterman algorithm.

4.5 Scoring matrices

In dynamic programming, the alignment uses scoring methods to quantify the probability of one residue being replaced by another in alignment; this method is known as the substitution matrix. The scoring method for nucleic acid sequence is relatively simple. A high or positive value is given for a match, and a negative value is assigned to a mismatch. Because amino acids with the same properties are more likely to match than those with similar properties, scoring for amino acids is quite complex due to the influence of physicochemical properties. Substitution of an amino acid with one having different physicochemical properties is disruptive to the role and structure of a peptide. For instance, the aromatic amino acids like tryptophan, tyrosine, and phenylalanine can easily substitute for each other without any disturbance because they all have the same chemical properties. The same happens with basic amino acids like lysine, arginine, and histidine. So, these are more frequent in substitution. But the amino acid cysteine, which can form a disulfide bond, is interrupted by enzymes and can destabilize protein structure; therefore, this kind of amino acid substitution does not occur frequently (Pearson, 2013).

4.5.1 Scoring matrices for amino acids

Substitution matrices for amino acids are 20*20. There are basically two kinds of matrices: first, interchangeable amino acid residues that are feature-dependent.

The second is dependent on empirical studies of amino acids. The first method is less precise than the second approach. The empirical method is the most favored and accepted method, and it consists of PAM and BLOSUM. These are thoroughly based on the statistics of substitution probability, and therefore, by analyzing the likelihood, they obtain a score for the particular alignment. If the one with high substitution scores higher, one of three things can happen: if the score is positive, it means that the frequency of residue substitution in homologous sequences is higher than when it happens randomly. If the score is 0, it represents the frequency that has occurred by chance as being equal to the frequency of substitution in homologous sequence data. If the score is negative, it means the frequency of substituting homologous sequence sets is less than the frequency expected by chance.

4.5.2 PAM (point accepted mutation)

PAM was first created by Margaret Dayhoff and his colleagues in the 1970s to develop substitution matrices of amino acids from related proteins having a minimum of 85% sequence identity. It was used for recreating ancestral groups of 34 superfamilies based on 71 pairs of sequences that were approximately 85% identical and based on 1572 variations. One unit of PAM is defined as 1% of the amino acid residue position that has changed, or one amino acid substitution per 100 amino acids. The family of PAM matrices includes PAM80, PAM120, and PAM250. The number indicates the basis of the matrix, which is the evolutionary distance between the pair of sequences. Because they have a greater evolutionary distance, the greater the digit, the greater the distance. The number of residue replacements used to calculate the PAM1 matrix To deduce the PAM-1 substitution matrix, a class of closely related strings with mutation frequencies corresponding to PAM-1 is selected. A phylogenetic tree can be constructed using the parsimony rule. When pam matrices are compiled, each mutation at the site is independent of any previous change at that site. PAM matrices are symmetrical because the direction of any change or mutation cannot be determined at a given region. It is derived from the global alignment method. It is a linear-space scoring method.

4.5.2.1 PAM score

1. It is a multi-step process that divides the mutational variation that occurred from the common ancestor for any residue by the total residue of amino acids in an alignment.
2. Normalizing expected frequencies of residues through random chance.
3. Known as the "log odd score" on a log scale.

4.5.2.2 Example

Assume C and D are two residues that randomly align; the probability that they aligned together randomly is Pr(C,D). The likelihood that C and D will meet is related to sequence alignment (orthologs: Po(C,D)).

Score function $= \log$ Po(C,D)/Pr(C,D).

Scoring matrix $= S(c,d) = log10(Pcd/PcPd)$.

By multiplying PAM 1 by itself, you can get other PAMs. For example, you can get PAM 80 by multiplying PAM 1 by itself 80 times, since PAM is defined as one mutation per 100 amino acid residues.

Limitation of PAM:

- The proteins studied had only a small amount of variation (85% identity), and alignment for more divergent sequences was not possible.
- Biased matrix due to its reliance on globular proteins.
- It is only based on one original dataset.

4.5.3 BLOcks SUbstitution matrix (BLOSUM)

This method was given by Henikoff and Henikoff in 1992 (Henikoff and Henikoff, 1992), and it overcomes the limitation of PAM matrices. It was developed by a group of around 2000 aligned, ungapped sites from 500 families of proteins and is known as the "BLOCKS" database. It is basically an ungapped alignment of highly conserved sites of protein. The presence of each residue pair in each column of all blocks is counted. Despite the extrapolation function, the actual percentage identity value of the sequence is used to construct the matrix in this case. It represents the sequence alignment with no more than 62% identity. It means that in a number system, BLOSUM is the opposite of PAM, and the lower the BLOSUM number, the more different or far apart a sequence is.

4.5.3.1 BLOSUM62

It represents the sequence alignment with no more than 62% identity. In the case of a numerical system, it means the opposite of PAM; the more divergent or distantly related a sequence is, the lower the BLOSUM number.

4.5.3.2 BLOSUM score

It is the log ratio of the substitution frequency of the observed amino acid residue to the expected probability of a specific residue. The log is taken to base 2, the final value is rounded off to the nearest integer, and then it is written in the substitution matrix.

Comparison between BLOSUM and PAM

More distant sequence

PAM 100 = BLOSUM90

PAM120 = BLOSUM80

PAM160 = BLOSUM60

PAM200 = BLOSUM52

PAM250 = BLOSUM45

4.6 **Word methods**

Though far more effective than dynamic programming, this method—also known as k-tuple methods—does not ensure that the optimum alignment solution will be discovered. These strategies are particularly helpful in large-scale database searches where a significant portion of candidate sequences are projected to share little in common with the intended set of query sequences.

The query is divided into a cassette of short, non-overlapping subsequences ("words"), which are then compared against database sequences. The offset is derived by subtracting the respective positions of the words in the two sequences being compared; if several different words provide the same offset, this indicates a region of alignment. These methods only use more sensitive alignment criteria if this area is found; as a result, many useless comparisons with sequences with no significant similarity are skipped. The FASTA and BLAST families of database search tools are well known for their use of word methods (Altschul et al., 1990).

Software for sequence alignment called FASTA is pronounced "fast A." David J. Lipman and William R. Pearson initially presented it in 1985. The software package's initial iteration was only intended to compare protein sequences. Programs for protein:protein, DNA:DNA, protein:translated DNA (including frameshifts), sorted or unordered peptide searches, among other capabilities, are included in the most recent edition. For figuring out whether or not something is statistically significant, the software has a more advanced way of shuffling.

In the FASTA technique, the user specifies a value k as the word length to search the database with. The technique is slower but sensitive for smaller values of k, making it appropriate for searches with a short query sequence.

The first method to employ a look-up table and the seed searching strategy for indexing the initial sequence is FASTA. The look-up table is used to find perfect match seeds of length "k" in both sequences at the start of the process. To locate all seeds across a diagonal between the two sequences, FASTA employs a "diagonal" approach. Different scores are given to seeds depending upon their position; seeds receive a positive score, and intermediate areas receive a negative score. So, clusters of high-scoring seeds add more to the score of local diagonals than clusters of low-scoring seeds. There may be "n" diagonal seeds produced throughout the procedure, of which FASTA stores the top 10. FASTA determines the maximum score "init1" as well as "good" diagonals. A diagonal with a score higher than the threshold is considered to be good. Then, by building a directed weighted graph around the seeds, all of these diagonals are integrated into a single, space-allowing, high scoring alignment. The optimum alignment is then identified and designated as "initn" on the most heavily weighted graph. The following phase of FASTA constructs a narrow band cantered along init1 with a width of "k." Next, FASTA uses the Smith-Waterman method to determine a local alignment that is best within this band.

The software primarily saves word-to-word matches of length k using a hash table pattern search. The software searches the returned word hits for segments containing a cluster of neighboring word hits. When the database is searched for all possible sequences, the software plots the scores of each sequence in a histogram and determines their statistical significance. The "E-value" quantifies the possibility that the alignment was caused by random chance alone. The value must be significantly less than 0.05. Very low E values in the results indicate that the sequences are homologs. A long list of steadily decreasing (E) values indicates a large sequence (gene, protein, or RNA) family. Long areas with intermediate similarity are more significant than small sections with high similarity.

FASTA has the drawback of only finding the region surrounding init1 while excluding the area that contributed to initn if the sequences in evaluation include several homology regions (two optimum diagonals). A seed, which is smaller than k, could be overlooked. The primary benefit above the optimum algorithm is its speed.

BLAST is one of the NCBI's tools. BLAST was designed to be a faster alternative to FASTA without sacrificing much accuracy. It uses a k-word search like FASTA, but instead of assessing every word match, it just evaluates the most significant ones. The default word length in most BLAST implementations is optimized for the query and database type, and it is only changed in exceptional instances, such as when searching with repetitive or very short query sequences. If the world length is approximately 3 amino acids or 11 bases, it would not require identical words, and if the words are similar, then there will be no alignment. BLAST makes similarity searches very quickly but also makes errors, e.g., it misses some important similarities and makes many incorrect matches. Two websites to discover implementations are EMBL FASTA and NCBI BLAST.

4.7 Multiple sequence alignment

Sequence alignment is the alignment of two or more sequences. The alignment of the sequences is the arrangement of two sequences in relation to one another. Alignment is the process of placing two sequences one after the other and looking for similarity regions and homology between them (Edgar and Batzoglou, 2006). For example, if there are two sequences that need to be aligned, one nucleic acid sequence (like ATCG) and another sequence (like TCA).

From Table 4.1, it can be deduced that there are three ways to align two sequences. We attempt to obtain the optimal alignment, which entails aligning two sequences in a manner that generates the greatest number of matches. Compare the homology or similarity between sequence 1 and sequence 2 in Table 4.2. Alignment is the process of obtaining matches. If there are more matches, then the sequences are more similar, and if there are fewer matches, then the sequences are less similar. Therefore, all feasible alignments of these two sequences are shown below.

There are no matches in Table 4.2, but there are three mismatches with one gap. We are searching for three rules, namely match, mismatch, and gap, to identify sequences. Since this sequence is short, there are fewer alignment options. However,

Table 4.1 First pair of three sequences.

A	T	C	G
T	T	C	A
Gap	Match	Match	Mismatch

Table 4.2 Second pair of the three sequences.

Sequence 1	A	T	C	G
Sequence 2	T	C	A	
	Mismatch	Mismatch	Mismatch	Gap

there are still many ways to align two sequences, such as when the software aligns 500–600 bp of nucleotides to align two sequences.

Consequently, there are two matches in Table 4.1 and one match in Table 4.3, but none in Table 4.2. As Table 4.1 received the highest number of matches, this indicates that it is the optimal alignment. When we upload data into a computer, it integrates all of the data and performs every conceivable alignment, looks for the marks distribution system, analyses the outcome, and then provides marks to each alignment. It then displays the alignment with the highest marks. When there is at least some resemblance between two sequences, they can be aligned. It will be harder to match two sequences from, say, plants and vertebrates. Therefore, the closely related sequences must be aligned.

Using multiple sequence alignment (MSA) is a crucial first step in any bioinformatics analysis of protein or nucleotide sequences since it allows researchers to determine how similar their many sample sequences are to one another. It is difficult to determine the evolutionary connection without using MSA, which is hence required for phylogenetic analysis. A phylogenetic tree cannot be constructed without it. In addition to this, it is also used for finding conserved domains (Gotoh, 1999).

4.7.1 Progressive alignment

In comparison to pairwise alignments, multiple sequence alignments are notoriously difficult to calculate. This method builds a hierarchical cluster of the sequences by

Table 4.3 Third pair of three sequences.

Sequence 1	A	T	C	G
Sequence 2	T		C	A
	Mismatch	Gap	Match	Mismatch

iteratively generating pairwise alignments. The sequence is first aligned using dynamics programming in progressive alignment. It builds an MSA starting with the most related sequences and gradually adds less related sequences or groups of sequences using the dynamic programming method (Taylor, 1986).

The problem with the progressive alignment method is its dependence on pairwise sequence alignment at the initial step for ultimate MSA results. Also, as this method is based on global alignment, it is not compatible with sequence alignments of different lengths. CLUSTAL W is perhaps the most suitable example of progressive alignment, along with MAFFT, T-COFFEE, and MUSCLE (Fig. 4.8).

4.7.2 Iterative method

Similar in operation to progressive approaches, iterative methods repeatedly realign the original sequences and add new sequences to the developing MSA. Since progressive techniques are always included in the final product, and once a sequence is aligned into the MSA, its alignment is not reevaluated, they are dependent on a high-quality initial alignment. While increasing efficiency, this approximation compromises accuracy. To optimize a generic objective function like producing a high-quality alignment score, iterative algorithms can, however, refer back to previously calculated pairwise alignments or subMSAs (Fig. 4.9).

The two software packages PRRN and PRRP are examples of iterative methods. They utilize a hill-climbing method to maximize its MSA alignment score and rectify alignment weights and "gappy" sections of the MSA repeatedly. PRRP performs better when revising an alignment created by a faster approach. MUSCLE (multiple sequence alignment by log-expectation), which is another well-known

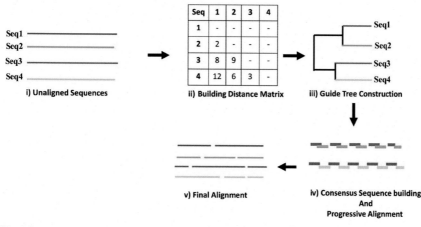

FIG. 4.8

Progressive alignment algorithm.

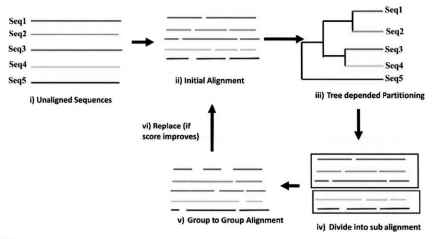

Seq1

Seq2

Seq3

Seq4

Seq5

i) Unaligned Sequences

ii) Initial Alignment

iii) Tree depended Partitioning

Seq1

Seq2

Seq3

Seq4

Seq5

vi) Replace (if score improves)

v) Group to Group Alignment

iv) Divide into sub alignment

FIG. 4.9

Iterative alignment algorithm.

iterative method, improves progressive methods by using a more accurate distance metric to figure out if two sequences are linked.

4.7.3 MSA filtering

The MSA approaches have several flaws because they rely on heuristic searches that are guided by faulty objective roies. As a result, even if their output is excellent overall, it is invariably polluted with faults. A number of software tools have been created with the goal of filtering MSAs so that only the most reliable parts remain. This is accomplished by removing sites, sequences, or remnants (by replacing them with the gap symbol, "−," or an ambiguity symbol, "?," "N," or "X," respectively).

It's crucial to make certain that filtering doesn't eliminate both the signal and the noise created by the improperly aligned sections. The equilibrium present between noise cutting and signal disappearance is critical in the case of MSA filtering.

Filtering techniques are mainly separated into two divisions. The very first one consists of ways for filtering MSA by completely deleting certain regions or sequences. These methods simply give the user two options for each region and sequence: "take it or leave it," which is why they are known as TILI-filtering techniques. MSA filtering techniques that function by hiding remnants (replacing them with the gap symbol "−" or a character depicting ambivalence such as "?," "N," or "X," as per the type of sequence) fall into the second group. These are termed "picky filtering" techniques because they grab bits of information from a region or sequence while ignoring the rest.

4.7.4 **Filtering techniques' fundamental principles**

A. Gaps show locations that are difficult to align and maybe saturated

Sequence alignment involves filling gaps. The number of gaps at a place directly affects how much effort the alignment approach must make; hence, the procedure is more likely to cause mistakes. Biologically, insertions and deletions are less prevalent in proteins than point substitutions. Several gaps show an aberrant evolutionary pattern, perhaps owing to an MSA problem. Multiple mutations at the same location are likely, obscuring the evolutionary signal.

B. Per site, a small number of residues with similar properties are expected

Homologous regions are likely to share traits, notably amino acid sequences. If all amino acids at one site are the same, they all descend from the same hereditary amino acid. Second, the protein has remained functional despite the physicochemical properties of the remnant at this point, and at least 19 substitutions have occurred to do this (which can suggest a site that is saturated). In this situation, removing this alignment piece would be safer. Hydrophobic or positively charged residues should be retained. Estimating residual transformation in one spot may be done in a variety of ways, from easy (a simple count of amino acids found on the site) to complicated.

C. Similarity in homologous sequences is expected

In most processes, homologous sequences are identified by their similarity. This ensures that the sequences that are supposed to be aligned have a minimum value of total similarity. Despite this, it's unusual for a portion of one or more sequences to deviate dramatically from the rest of the alignment. This sequence is unlikely to be homologous to the remaining area, maybe because it was not properly aligned and the segment is homologous to a different portion of the alignment or because it has no homology at all. This latter circumstance can be recognized before sequences are aligned, and various approaches have been developed. This filtering should be done before sequence alignment, since long insertions in particular sequences might hamper the MSA.

D. Orthologous sequences should be consistent across loci (post-filtering)

Alignment filtering algorithms will almost certainly fail to detect a non-orthologous sequence that is homologous with the remaining sequences. However, in the case of a phylogenomic setting, it is conceivable to inspect the MSAs for every locus under consideration. With a high count regarding taxa and genes, it is easier to learn loci, along with the speed of evolution of taxonomic categories. The OrthoMaM v10 process has one basic technique for detecting those sequences that are non-orthologous, whereas Phylo-MCOA offers a more detailed solution. However, this is not applicable in cases where the study is focused on the evolution of an entire family of genes (both paralogous and orthologous).

4.7.5 Programs and methods for multiple sequence alignment

4.7.5.1 Clustal family

MSA's most well-known program is the Clustal family. There are two programmes in this family: ClustalX and ClustalW. Clustal family tools are very fast and relatively reliable; they work in the same way as any other progressive alignment tool by either using PAM250 or BLOSUM62 with global DP for the pairwise alignment of all the provided sequences.

The pairwise scores are then used by Cluster to generate a neighbor joining tree; some early versions are used to follow the UPGMA method. The sequence alignment is done by following the leaves inward on the tree. Clustal is much more sensitive than other alignment tools as it considers the gap penalties and the short hydrophilic stretches as an indication of a random coil region so that the gap penalties are reduced for these stretches.

Clustal Omega: The most recent and best-performing webserver in the Clustal family. It can align around 200,000 sequences with just a single run and give results in a few hours. In terms of accuracy, it's very similar to other high-quality web servers, but on a large scale of sequencing, Clustal Omega is quite good in terms of both quality and the completion time of alignment. The very first step that this server follows is pairwise alignment by the k-tuple method. Afterward, the mBed method is used for clustering along with the K-means clustering method. Then it is followed by the building of a guide tree with the help of the UPGMA method. At last, the HHalign package aligns two hidden Markov model profiles and generates the alignment of multiple sequences.

4.7.5.2 DIAlign

DIAlign combines global and local pairwise alignments. DIAlign has MSA, which is composed of segments of equal length that have statistical similarity, which is quite significant. DIAlign is very similar to FASTA alignment.

4.7.5.3 Tree-based consistency objective function for alignment evaluation (T-coffee)

It is also a progressive multiple sequence alignment tool that uses ClustalW and LALIGN at the start to obtain a library of primary pairwise sequences that combine global pairwise sequences from ClustalW and local pairwise sequences from LALIGN. T-Coffee creates an extended library based on primary library triplets, where the third sequence aligns the other two sequences and creates a new pairwise alignment; duplicate pairs are removed completely. This extended library is actually a list of weighted residue pairs. From this extended library, the final multiple sequence alignment is generated by performing progressive alignment. 3D-coffee is a new and improved version of T-coffee. It uses a specialized server to create a 3D structure alignment, which shows a superposition, which is an area where two structures overlap. This superposition of two sequences is used to increase the weight of residues in the library (Notredame et al., 2000).

4.7.5.4 FAlign

FAlign uses a combination of both iterative and progressive algorithms to align multiple sequences. To use FAlign, users need to identify and define all the regions in all the sequences where a motif is present. At motif boundaries, all the sequences split, and the obtained segments are then aligned progressively. These segments are aligned, and they are properly assembled, which generates an alignment. BLOSUM62 is used in FAlign to create a score matrix that gives a sum-of-pair score. The alignment score is improved by the non-motif and by shifting the gaps iteratively and randomly.

4.7.6 Representation and structural inference

To generate multiple sequence alignments, we must first select more than two sequences from which to align and find similarity data. So after retrieving the sequences from FASTA, they are copied to the Clustal Omega web server box and run for the results. As it is very rapid, it will generate results within a few seconds. Using two examples, we will learn about the representation and structural inference of multiple sequence alignment.

Example 1: Sequences of different species are retrieved to use as input in clustal. The species were: panther histone, leopard mitochondria, tiger mitochondria, cat cloning vector, and synthetic cat (Fig. 4.10).

Because the sequences are from different species and have less similarity, there are no common matches between them. See any colon, semicolon, or star presence here that states the presence of matches or is very close to matches because the 5 sequences are very different from each other.

Example 2: Another five sequences of different organisms, but from the same class of species, Synthetic Cat, Cat Cloning Vector ThyA-Cat, Cat Expression Vector, and Cat Plasmid are the organisms (Fig. 4.11).

From the result, it is observed that there is alignment in the sequence. There are semicolons present that represent that they are much closer to the matches, and only 1 residue is different. Following that, there are colons, indicating that they are also close to the matches by the difference of two different residues.

4.8 Phylogenetics

Phylogenetic inference methods use heritable traits like DNA, amino acid sequences, or morphology to suggest links based on a trait evolution model. The outcome is a phylogeny (also known as a phylogenetic tree), a diagram demonstrating evolutionary links between species. Rooted and unrooted phylogenetic trees are the two forms of phylogenetic trees. A rooted tree diagram shows the tree's ancestry. An unrooted tree diagram (a network) has no assumptions about the ancestral line; therefore, it doesn't illustrate the origin or "root" of the species or the direction of anticipated evolutionary transitions.

```
CLUSTAL O(1.2.4) multiple sequence alignment


Cat             --------MGPSVRPGFLVVV---------IGLQFVAAS-----MEVNSRKFEPMVAFIC 38
Panther         MGPASPAARGLSRRPGQPPLPLLLPLLLLLLRAQPAIGSLAGGSPGAAEAPGSAQVAGLC 60
Synthetic_cat   -----------------MLHIIMPSMTFLLAKG--FFFFQVPTDGNAGLLAEPQIAMFC 40
Leopard         --------------------MLPSLALLLLAAWTVRALEVPTDGNAGLLAEPQIAMFC 38
Tiger           --------------------MLPGLALLLLAAWTARALEVPTDGNAGLLAEPQIAMFC 38
                                                      :    . :* .* :*


Cat             NKPAMHRV--PSGWVPDDDPAKSCVKQPEEILEYCKKLYPDHDITNVLQASYKVTIPNWC 96
Panther         GRLTLHRDLRTGRWEPDPQRSRRCLRDPQRVLEYCRQMYPELQIARVEQATQAIPMERWC 120
Synthetic_cat   GRLNMHMNVQNGKWDSDPSGTKTLSP------------LRWLTITNVVEANQPVTIQNWC 88
Leopard         GKLNMHMNVQNGKWESDPSGTKTCIGTKEGILQYCQEVYPELQITNVVEANQPVTIQNWC 98
Tiger           GRLNMHMNVQNGKWDSDPSGTKTCIDTKEGILQYCQEVYPELQITNVVEANQPVTIQNWC 98
                .: :*   .  * * .. ::                 *:.* :*.  : : .**


Cat             GFNVTHCHKHGNHTVRPFRCLVGPFQSEALLVPEHCIFDHYHDPRVCNEFDQCNETAMSK 156
Panther         GGSRSGSCAHPHHQVVPFRCLPGEFVSEALLVPEGCRFLHQERMDQCESSTRRHQEAQEA 180
Synthetic_cat   KRGRKQ-CKTHPHFVIPYRCLVGEFVSDALLVPDKCKFLHQERMDVCETHLHWHTVAKET 147
Leopard         KRGRKQ-CKTHTHIVIPYRCLVGEFVSDALLVPDKCKFLHQERMDVCETHLHWHTVAKET 157
Tiger           KRDRKQ-CKTHPHIVIPYRCLVGEFVSDALLVPDKCKFLHQERMDVCETHLHWHTVAKET 157
                       . .       * * *:*** * * *:*****. * * *.    *:   :: * .


Cat             CSARGMTTQSFAMLWPCQEPGHFSGVEFVCCPKVSLIPESTEAPKSSPP---------- 205
Panther         CSSQGLILHGSGMLLPCGS-DRFRGVEYVCCPPPGTPDPSGTAVG-DPSTRSWPPGSRVE 238
Synthetic_cat   CSEKSTNLHDYGMLLPCGI-DKFRGVEFVCCPLAEESDNVDSADAEEDDSDVWWGGADTD 206
Leopard         CSEKSTNLHDYGMLLPCGI-DKFRGVEFVCCPLAEESDSIDSADAEEDDSDVWWGGADTD 216
Tiger           CSEKSTNLHDYGMLLPCGI-DKFRGVEFVCCPLAEESDHVDSADAEEDDSDVWWGGADTD 216
                ** :.   :. .** **   .:* ***:****         *   .
```

FIG. 4.10

MSA 5 sequences of different species.

```
CLUSTAL O(1.2.4) multiple sequence alignment


Cat_Plasmid            MCLLIPWFQITLLLLWIYLFFYPSLLSLFLVPTDGNAGLLAEPQVAMFCGKLNMHMNVQN 60
Synthetic_Cat          ---MLPALALVLLAAWT-------ARALEVPTDGNAGLLAEPQVAMFCGKLNMHMNVQN 49
ThyA-Cat               -----------------------------------------MFCGKLNMHMNVQN 14
Cat_Cloning_Vector     ---MLPALALVLLAAWT-------ARALEVPTDGNAGLLAEPQVAMFCGKLNMHMNVQN 49
Cat_Expression_Vector  ---MLPALALVLLAAWT-------ARALEVPTDGNAGLLAEPQVAMFCGKLNMHMNVQN 49
                                                                 **************


Cat_Plasmid            GKWESDPSGTKTCIGTKEDILQYCQEVYPELQITNVVEANQPVTIQNWCKRGHKQCKTHA 120
Synthetic_Cat          GKWESDPSGTKTCIGTKEDILQYCQEVYPELQITNVVEANQPVTIQNWCKRGHKQCKTHA 109
ThyA-Cat               GKWESDPSGTKTCIGTKEDILQYCQEVYPELQITNVVEANQPVTIQNWCKRGHKQCKTHA 74
Cat_Cloning_Vector     GKWESDPSGTKTCIGTKEDILQYCQEVYPELQITNVVEANQPVTIQNWCKRGHKQCKTHA 109
Cat_Expression_Vector  GKWESDPSGTKTCIGTKEDILQYCQEVYPELQITNVVEANQPVTIQNWCKRGHKQCKTHA 109
                       ************************************************************


Cat_Plasmid            RIVIPYRCLVGEFVSDALLVPDKCKFLHQERMDVCETHLHWHTVAKGGPTCSEKSTSLHD 180
Synthetic_Cat          RIVIPYRCLVGEFVSDALLVPDKCKFLHQERMDVCETHLHWHTVAK--ETCSEKSTSLHD 167
ThyA-Cat               RIVIPYRCLVGEFVSDALLVPDKCKFLHQERMDVCETHLHWHTVAK--ETCSEKSTSLHD 132
Cat_Cloning_Vector     RIVIPYRCLVGEFVSDALLVPDKCKFLHQERMDVCETHLHWHTVAK--ETCSEKSTSLHD 167
Cat_Expression_Vector  RIVIPYRCLVGEFVSDALLVPDKCKFLHQERMDVCETHLHWHTVAK--ETCSEKSTSLHD 167
                       *********************************************   ***********


Cat_Plasmid            YGMLLPCGIDKFRGVEFVCCPLAEESDNIDSADAEEDDSDVWWGGADADYADGSEDKVVE 240
Synthetic_Cat          YGMLLPCGIDKFRGVEFVCCPLAEESDNIDSADAEEDDSDVWWGGADADYADGSEDKVVE 227
ThyA-Cat               YGMLLPCGIDKFRGVEFVCCPLAEESDNIDSADAEEDDSDVWWGGADADYADGSEDKVVE 192
Cat_Cloning_Vector     YGMLLPCGIDKFRGVEFVCCPLAEESDNIDSADAEEDDSDVWWGGADADYADGSEDKVVE 227
Cat_Expression_Vector  YGMLLPCGIDKFRGVEFVCCPLAEESDNIDSADAEEDDSDVWWGGADADYADGSEDKVVE 227
                       ************************************************************
```

FIG. 4.11

MSA results.

Phylogenetic studies are used not just for their intended purpose of deducing phylogenetic patterns and taxa but also for the purpose of depicting connections among individual organisms or gene copies.

4.8.1 Molecular phylogenetics

The technique of inferring evolutionary relationships that develop as a result of molecular evolution and generating a phylogenetic tree is known as molecular phylogenetics. In molecular phylogenetics, genetic and hereditary molecular distinctions, especially DNA sequences, are analyzed to uncover an organism's evolutionary origins. A subset of molecular systematics, which uses molecular data in taxonomy and biogeography, is molecular phylogenetics. The fields of molecular phylogenetics and evolutionary biology are inextricably connected.

4.8.2 Phylogenetics trees

A phylogenetic tree, also known as an evolutionary tree or phylogeny, is a branching diagram that shows the evolutionary relationships between biological species or other entities based on physical or genetic similarities and differences. The lengths of the edges in some trees may be interpreted as estimations of the passage of time, and each node in a rooted phylogenetic tree represents the estimated most recent common ancestor. A taxonomic unit is a name that is assigned to each node in the tree. Internal nodes are sometimes called hypothetical taxonomic units because they are not easy to see.

The ancient idea of a hierarchical ladder connecting all forms of life gave rise to the modern "tree of life" metaphor. Edward Hitchcock's "paleontological chart" in his Elementary Geology book is one of the earliest examples of a "branching" phylogenetic tree because it shows the geological links between different species of plants and animals.

In his landmark essay The Origin of Species, published in 1859, Charles Darwin produced one of the earliest depictions and promoted the notion of an evolutionary "tree." Even though they have been around for almost a century, tree diagrams are still widely used by evolutionary biologists to convey the idea that speciation happens through adaptive and semi-random branch splitting. Over time, species classification has grown increasingly fluid and less rigid.

4.8.3 Properties

A rooted phylogenetic tree is a directed tree that has one node at the root that corresponds to the (often accepted) most recent common ancestor of all the organisms at the tree's leaves. Despite the fact that it has no parents, the root node in the tree is the parent of all other nodes. As a result, the root is a node of degree 2, whereas the remainder of the internal nodes are all at least nodes of degree 3. (The term "degree" refers to the sum of all incoming and outgoing edges.)

Unrooted trees do not assume any ancestry and merely illustrate the relationship between the leaf nodes. If a tree's root is absent, it may always be created from an existing, rooted tree. On the other hand, determining an ancestor's lineage is necessary to infer a tree's root. Common ways to do this are to include an "outgroup" in the input data, force the root to be between the "outgroup" and the other species in the tree, or make extra assumptions about how fast each branch is changing.

Bifurcating and multifurcating branches are possible in both rooted and unrooted trees. Unrooted bifurcating trees have precisely three neighbors at each internal node, while rooted bifurcating trees (which form a binary tree) have precisely two descendants emanating from each inner node. On the other hand, a multifurcating, unrooted tree may have more than three neighbors. Trees can be labeled or unlabeled. Labeled trees have values assigned to the nodes, but an unlabeled tree, also known as a tree form, only depicts the structure of the tree (Kapli et al., 2020).

4.8.3.1 Phylogenetic trees and networks

Depending on the input data and technique used, computationally produced phylogenetic trees might be rooted or unrooted. Evolutionary processes like hybridization and horizontal gene transfer may be simulated by connecting rooted and unrooted phylogenetic trees to form rooted and unrooted phylogenetic networks. Dendrograms, phylograms, cladograms, and Dahlgren diagrams are only a few examples of the many tree representations that exist.

4.8.4 Building methods

Computational phylogenetics methods are used to generate phylogenetic trees from a large number of input sequences. Distance-matrix methods, such as neighbor-joining or UPGMA, which calculate genetic distance from multiple sequence alignments, are the easiest to construct. These methods aren't based on any sort of evolutionary theory. Simpler tree-building algorithms are used by several sequence alignment tools, including ClustalW. Maximum parsimony is another simple way to figure out phylogenetic trees, but it does require an implicit model of how evolution works.

4.8.5 Distance matrix method

To classify sequences, distance-matrix methods depend on a metric called "genetic distance." The percentage of mismatches at aligned sites is often used to calculate distance, with gaps either being disregarded or counted as mismatches. The goal of distance techniques is to create an all-to-all matrix that represents the distance between each pair of sequences in the sequence query set. A phylogenetic tree is made when sequences that are very similar to each other are put under the same inner node and the lengths of the branches are close to the distances between the sequences.

4.8.5.1 UPGMA

The UPGMA (Unweighted Pair Group Method with Arithmetic Mean) approach produces rooted trees and requires a constant-rate assumption, which necessitates an ultrametric tree with equal distances between root and branch tip. It is the oldest and simplest distance matrix method. It uses a sequential clustering algorithm and clusters the two closest species. The process is repeated until all species are grouped. The main advantage of this approach is that it is fast and is suitable for analyzing large data sets. There are several models available with many parameters in UPGMA that improve the distance estimation. However, this method lacks in some aspects, as a lot of information is lost, and the history of sites can only be investigated through character-based analyses.

4.8.5.2 Neighbor-joining

Neighbor-joining approaches extend general cluster analysis techniques to sequence analysis by using genetic distance as a grouping parameter. It is a method of clustering that works from the bottom up, and the term "neighbors" refers to taxa that are related to one another by a single node in an unrooted tree. The simple neighbor-joining method produces unrooted trees, but it does not assume that lineages evolve at the same pace. Up until the tree is resolved, the closest neighbors are sequentially linked by a new node. Even for many hundreds of sequences, the NJ technique is quick. However, the approach lacks accuracy because there is no attempt to correct for potential bias. NJ also does not examine all possible topologies.

4.8.5.3 Fitch Margoliash method

The Fitch-Margoliash method groups people based on their genetic distance by using a weighted least squares algorithm. To account for the higher inaccuracy in calculating distances between distantly related sequences, the tree-building strategy gives greater weight to tightly linked sequences. To prevent significant artifacts when computing relationships between closely and distantly related groups, the distances employed as the method's input should be normalized. According to the distance linearity requirement, the anticipated value of the sum of two branch distances must be equal to the expected value of the two branch length sums. A substitution matrix, such as the one produced by the Jukes-Cantor model of DNA evolution, is used to make this correction. But the distance has to be changed when the branch rates start to change at different speeds.

Another variant of the strategy has been proven to increase the algorithm's effectiveness and robustness, especially in cases involving relatively small distances. The least-squares criteria for distance measurement is more precise than neighbor-joining approaches but less time-efficient. At a higher processing cost, a further improvement can be added to take into account correlations between distances that come from a number of closely related sequences in the data set.

4.8.5.4 Maximum parsimony

Maximum parsimony (MP) is a method for deciding which phylogenetic tree best reflects the available sequencing data and needs the fewest overall evolutionary events. In this approach, sequences are aligned in order to build a tree that reduces the total number of mutations while simultaneously reducing the sum of all branch lengths. It aligns the sequences directly without making use of a distance matrix or an evolutionary model, and it entirely overlooks the notion that there may be numerous mutations. It was first designed to work with morphological characters. The fundamental idea behind MP is that it estimates the fewest possible replacements for any given topology. Among the benefits of this method are: (a) It is a simple method that doesn't make any assumptions and doesn't depend on any evolution model. However, if the substitution rates vary extensively between lineages, MP can also create incorrect topologies.

4.8.5.5 Maximum likelihood

The maximum likelihood approach assigns probabilities to fictitious evolutionary trees by using standard statistical methods for deriving probability distributions. The method calls for a substitution framework to calculate the probability of certain mutations; typically, a tree is judged to have a lower probability if it takes more mutations at inner nodes to account for the observed phylogeny. Maximum likelihood permits variable evolution rates across locales and lineages. As a result, this approach offers more statistical flexibility than maximum parsimony.

The "pruning" technique is frequently used to narrow the search space by more accurately predicting the likelihood of subtrees. The approach works backward from a node whose only descendants are leaves to the "bottom" node, estimating the probability for each site in nested sets in a linear manner. The procedure will only result in the production of rooted trees if the replacement model is irreversible, which isn't usually the case in biology.

This approach offers the most compelling theoretical justification for its use. Experiments using sequence simulation have demonstrated that this approach is superior to all others in terms of effectiveness in the majority of scenarios. However, this strategy relies heavily on the use of computers. Because there are so many different trees to consider, it is almost never possible to examine them all. A preliminary investigation of potential tree locations is carried out. The mathematical certainty of generating the tree with the highest probability has been lost.

4.8.6 Bayesian inference

In a manner similar to maximum likelihood methods, phylogenetic trees may be constructed using Bayesian inference. Bayesian approaches estimate a prior probability distribution of the potential trees. This can be as simple as the probability of any one tree out of all the possible trees that can be made from the data, or it can be a more complicated estimate based on the idea that divergence events like speciation are random. The prior distribution that is used in Bayesian inference phylogenetics is seen differently by different users.

Examples of Bayesian phylogenetics choices include randomly swapping the descendant subtrees of a random internal node between two interrelated trees and performing a circular permutation of the leaf nodes of a postulated tree at every phase. The use of Bayesian approaches in phylogenetics has been met with much debate as a result of a lack of transparency in previously published research about the decision-making process for the prior distribution, acceptance criterion, and move set.

Bayesian procedures use the posterior distribution to reconstruct a tree that depicts the most likely clades, whereas likelihood approaches locate the tree with the highest probability given the data. It is possible, however, for estimates of clade posterior probability to be quite inaccurate, especially for clades that aren't extremely probable. This has led to the introduction of several strategies for estimating the posterior probability.

References

Altschul, S.F., Gish, W., Miller, W., Myers, E.W., Lipman, D.J., 1990. Basic local alignment search tool. J. Mol. Biol. 215, 403–410. https://doi.org/10.1016/S0022-2836(05)80360-2.

Edgar, R., Batzoglou, S., 2006. Multiple sequence alignment. Curr. Opin. Struct. Biol. 16.

Gotoh, O., 1999. Multiple sequence alignment: Algorithms and applications. Adv. Biophys. 36.

Henikoff, S., Henikoff, J.G., 1992. Amino acid substitution matrices from protein blocks. Proc. Natl. Acad. Sci. U. S. A. 89, 10915–10919. https://doi.org/10.1073/PNAS.89.22.10915.

Kapli, P., Yang, Z., Telford, M.J., 2020. Phylogenetic tree building in the genomic age. Nat. Rev. Genet. 21 (7), 428–444. https://doi.org/10.1038/s41576-020-0233-0.

Mullan, L., 2006. Pairwise sequence alignment - it's all about us! Briefings Bioinf. 7, 113–115. https://doi.org/10.1093/BIB/BBK008.

Naghibzadeh, M., Babaei, S., Behkmal, B., Hatami, M., 2021. Divide and conquer approach to long genomic sequence alignment. In: ICCKE 2021 - 11th International Conference on Computer Engineering and Knowledge, pp. 399–405. https://doi.org/10.1109/ICCKE54056.2021.9721501.

Needleman, S.B., Wunsch, C.D., 1970. A general method applicable to the search for similarities in the amino acid sequence of two proteins. J. Mol. Biol. 48, 443–453. https://doi.org/10.1016/0022-2836(70)90057-4.

Notredame, C., Higgins, D.G., Heringa, J., 2000. T-coffee: A novel method for fast and accurate multiple sequence alignment. J. Mol. Biol. 302, 205–217. https://doi.org/10.1006/jmbi.2000.4042.

Pearson, W.R., 2013. Selecting the right similarity-scoring matrix. Curr. Protoc. Bioinformatics 43, 3.5.1–3.5.9. https://doi.org/10.1002/0471250953.BI0305S43.

Rice, P., Longden, L., Bleasby, A., 2000. EMBOSS: The European molecular biology open software suite. Trends Genet. 16, 276–277. https://doi.org/10.1016/S0168-9525(00)02024-2.

Saloom, R.H., Khafaji, H.K., Tobia, R.H., 2022. A survey for the methods of detection and classification of genetic mutations. Indones. J. Electr. Eng. Comput. Sci. 28, 1796—1816. https://doi.org/10.11591/IJEECS.V28.I3.PP1796-1816.

Smite, T.F., Waterman, M.S., 1981. Identification of common molecular subsequences. J. Mol. Biol. 147, 195—197.

Sun, J., Chen, K., Hao, Z., 2018. Pairwise alignment for very long nucleic acid sequences. Biochem. Biophys. Res. Commun. 502, 313—317. https://doi.org/10.1016/J.BBRC.2018.05.134.

Taylor, W., 1986. Identification of protein sequence homology by consensus template alignment. J. Mol. Biol. 188.

Genetic variations

5.1 Introduction

Human genetics is the study of the species that the majority of us regard as being the most significant on planet Earth: *Homo sapiens*. The desire to have a deeper understanding of ourselves is one motive for researching human genetics. The past 20 years have experienced robust human genetic variation. By understanding the genetic variation of an individual or a group, researchers are able to understand the organism and how they are able to adapt in a challenging environment. It was also observed that most of the variations are useless, which means the variation does not affect the ability to survive or adapt. Not all variations are positive or neutral; some of them have a negative effect on DNA sequence, which can cause diseases like Huntington's disease and cystic fibrosis. Finding out how genetic factors contribute to many human diseases is one of the advantages of studying genetic variation in humans.

The variations also contribute to molecular genetics, biology, and the biomedical field, which have exploded due to the number of variations in DNA sequence and genotypes that have been spawned. The human genetic data has increased the understanding of genetic variation patterns that occur in an individual or group, apart from the online database that is being provided in which genome variants are being published.

Homo sapiens hasn't had nearly as much time to build genetic diversity as most other species on the planet, which mostly predate mankind by vast epochs. The amount of genetic variation—biochemical individuality—between any two humans is approximately 0.1%. Any two (diploid) individuals have approximately 6×10 different bp, which is a significant factor in developing automated processes for analyzing genetic variation.

Most of the time, genetic variation is a natural thing that happens when DNA replication goes wrong. The term "variation" means a change in the genetic sequence. Due to genetic variations, all humans are different from each other in terms of hair, color, height, or even the structure of our faces. A group of individuals from the same species with similar characteristics have differences, these differences are due to variations. SNPs are the most prevalent form of variation in humans. The specialty of SNPs is that only one/single nucleotide is being replaced by the other

All About Bioinformatics. https://doi.org/10.1016/B978-0-443-15250-4.00002-2

nucleotide in the whole DNA sequence. SNPs are dispersed naturally throughout the DNA of each individual. Approximately 4–5 million SNPs exist in a person's genome, since they appear around once every 1000 bases on average. A variant is deemed to be an SNP if it is present in approximately 1% of the population. Scientists have detected more than 600 million SNPs in global populations. The majority of SNPs do not impact development or health.

Nonetheless, a number of these genetic differences have been demonstrated to be crucial for human health research. SNPs assist in predicting a patient's reaction to certain treatments, vulnerability to environmental variables like pollution, and susceptibility to disease. It also helps find out how genetic changes linked to illness are passed down through families. These variations can be observed in genes, DNA, proteins, chromosomes, and even in the function of proteins. Although all the individuals have 99.9% similar genomes and only 0.1% of the variation in the genome, we are all different phenotypically. Thus, this small amount of variation accounts for such a significant difference. This difference may affect how an individual reacts to a particular drug or a particular environment. Thus, it is necessary to understand and evaluate the genotype-phenotype associations. Understanding such associations also helps us to devise therapies according to the patient's genetic makeup to provide a better therapeutic effect. Bioinformatics, or computational biology, has aided in pharmacogenomics and personalized medicine due to the integration of information available through whole genome sequencing and clinical studies. Several databases have contributed to the development of the field. These databases include PharmGKB, DrugBank, OMIM, and other such databases (Lek et al., 2016).

5.2 Types of variations

In humans, genetic variation can take many forms, some of which occur naturally and others which are influenced by environmental factors. Changes in single nucleotides, addition or deletion that causes a change in copy number, which ultimately changes the copy number variation (CNV), tandem repeats, recombination in homologous chromosomes, i.e., translocations and inversions (also known as copy neutral variations) are all examples of variation. The size of variation ranges from a single nucleotide to a megabase. Restriction fragment length polymorphism (RFLPs), single nucleotide polymorphism (SNPs), and indel are all types of single nucleotide substitutions or alterations in which a single nucleotide is changed or replaced from a specific locus in a DNA sequence (Eichler, 2019).

The chromosome microarray test is done to find out if there are any missing or extra chromosomes. Any piece of chromosome gained or lost is termed a "copy number variant." All humans possess CNVs, most of which are harmless and do not cause any type of disorder. CNVs are a natural part of the evolution process. If the CNV in question is small and not linked to any important genes involved in the disorder, it is considered harmless. Also, CNV inherited from healthy parents is likely to be harmless. However, if the size of the CNV is large and contains an important gene, it can be very harmful. But if the CNV is initiated in the offspring, then it can be very dangerous, and the person faces intellectual disabilities and multiple congenital syndromes (Fig. 5.1).

The next type of variation is the SNPs. Each SNP have four allele. The SNPs are very abundant in number which can help in identifying the variation in the sequence. SNP is based on the oligonucleotide hybridization analysis which makes the detection of SNPs very fast and very precise.

Fig. 5.2 shows variations like single nucleotide changes, in which only one nucleotide is altered or replaced by another single nucleotide. GTAGGCCATGCA has been taken as the reference DNA sequence for all the variations. In the first case, single nucleotide substitution has been explained, in which the single nucleotide cytosine is being substituted by adenine. In this, insertion and deletion of single nucleotides are being observed. The second type of variation is tandem repeats; these repeats are short sequences of DNA that repeat themselves in whole. The variable number of tandems that repeat represents the total number of repeats in the whole sequence. The reason for insertion or deletion at a particular position is unknown.

5.3 **Effects of genetic variation**

The main effect of any kind of genetic variation is to give a population the flexibility to deal with environmental challenges. Previous experience has shown that genetic variation always helps and boosts the organism's chances of adapting to any type of challenging environment and prepares the organism for any type of unexpected situation. The organism reproduces to give birth to its offspring. The formation of offspring of the same kind is the primary reason for reproduction. However, it is not possible, to have identical offspring just like the parents, except in the case of clones, which are genetically identical to their parents. But in almost all cases, offspring are somewhat different from their parents. It is inevitable that there will be a difference between the parents and the offspring. Variations refer to the

FIGURE 5.1

Characteristics of variations being harmful and harmless.

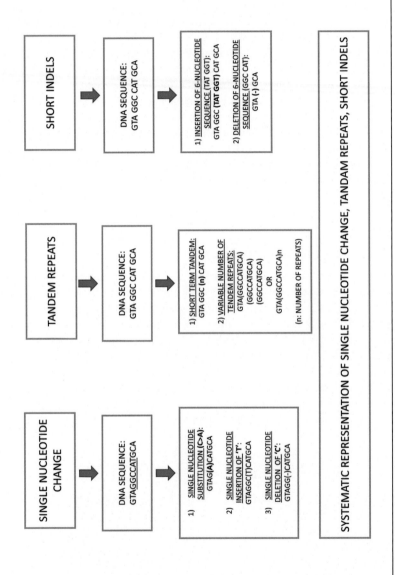

FIGURE 5.2

Systematic representation of single nucleotide change, tandem repeats, and short indels.

differences that exist between parents and their offspring. And these variations, over time, cause evolution. Evolution refers to any change in a population that takes years and improves organism capabilities. It is a very slow process and occurs in populations depending on environmental factors. The effect of any type of genetic variation is linked with evolution; the genetic variation is directly proportional to evolution. The more genetic variation, the more evolution; the less genetic variation, the lower the chances of evolution (Nei, 1983) (Fig. 5.3).

But still, not all genetic variation causes' evolution; the variation that will be passed on to the next generation causes evolution in the population and is called hereditary variation. These variations are observed in the egg or sperm cell so as to pass on the trait to the next generation. The gene variant causes permanent changes in the sequence of DNA, which will further form the gene. In this type of genetic variation, it is not mandatory that it will cause a disorder; thus, the "gene variant" term is given to the permanent changes in the DNA. Gene variants are different from single nucleotide polymorphism because they can change more than one nucleotide. In single nucleotide polymorphism, only one nucleotide changes.

The variants can be passed from parents to offspring and occur over the course of an individual's lifetime. If the gene variant is inherited, it is called a hereditary variant. The variation occurs in the germ cells, i.e., the egg or sperm cells of the parents, and is hence also considered a germline variation. When the egg and sperm cells containing variants form an embryo; the cells of this embryo will contain all the variant DNA from the parents and carry those variants through the rest of their lives. To function effectively, each cell requires hundreds of proteins to do their assigned jobs at the appropriate times and locations. Gene variations (also known as mutations) can occasionally compromise the function of one or more proteins. A variant can cause a protein to malfunction or not be formed at all by modifying the gene's instructions for protein synthesis. When a variation affects vitally essential proteins, it may interfere with normal growth or produce a genetic condition, generally known as a disease caused by variations in one or more genes. Genes do not cause disease by themselves. Instead, genetic diseases are caused by changes that change or take away a gene's function.

For instance, when individuals refer to "the gene of cystic fibrosis" they are often referring to a mutation of the CFTR gene that causes the disease. Every individual, including those without cystic fibrosis, has a variant of the CFTR gene (Studer et al., 2013).

5.4 Biological database

For the advancement of humanity, biology now incorporates numerous technological aspects. This expansion of biology has generated numerous data-rich outcomes that require compilation and organization. Databases are significant tools for scientists to use in analyzing and explaining a wide range of biological phenomena, from the structure of biomolecules and their interactions to the entire metabolism of

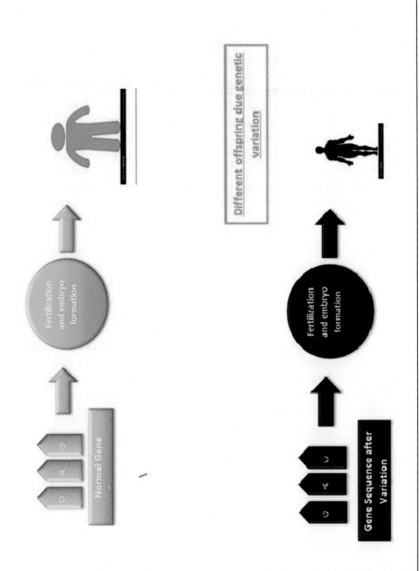

FIGURE 5.3

Different offspring's due to genetic variation.

organisms and comprehending species evolution. This knowledge aids in the fight against diseases, the development of drugs, the prediction of specific hereditary diseases, and the discovery of fundamental links between species throughout the history of life. A database of genetic variants offers information about genetic variances. Researchers submit data to these databases, which collect, organize, and publicly document evidence linking a human genetic variant to a disease or condition. Based on the current level of knowledge, these databases may make statements regarding genetic variations, Understanding the relationships between genotypes (an organism's genetic code) and diseases or disorders can help in the diagnosis and treatment of people with genetic problems. A few of the genetic variant's databases are explained below:

5.4.1 Database of human genetic variation

There are various types of biological databases that help in collecting the data of various experiment results for further purposes. One of the biological databases is the Human Genome Variation Database (HGV) (Fredman et al., 2002). Human Genome Variation maintains a publicly available online collection of genetic variation as documented in published research reports. Following evaluation by the editors, a standard range of data about every variation is retrieved from journal-published articles and added to the library. It's a free database that may be accessed and utilized by anyone.

The search bar provides filters like region, mutation type, and zygosity type, which makes the search more accessible and easier. The HGVD website allows users to view the number of samples, genotype and allele frequencies, coverages, and eQTL (expression QTL) significances. The key point of HGV database is that it provides data reports which have all the short reports regarding the human genome variation. It also has reports regarding the variability, which has the main role of describing the disease causing "genomic variants" and also specifies the frequency of those variants. The data report and HGV database are interlinked with each other. Every piece of information in the HGV database is organized into "sets" that are connected to their corresponding data report, which will have a connection to the corresponding information in the database. The HGV database collects different types of data, involving OMIM data, de novo/inherited, name of the gene, chromosome, protein alteration, mutation type, GenBank accession number, disease, phenotype, region, codon base change, genome position, and references (Li et al., 2012) (Fig. 5.4).

ClinVar is a publicly available database of research detailing the associations between phenotypes and human variants, along with supporting documentation. It provides access to and facilitates discussion regarding the stated links between human variation and recorded health status, as well as the background of this assessment. It also handles submissions describing variations detected in patients' specimens, clinical significance claims, submitter information, and any supporting data. As per the HGVS guideline, the alleles mentioned in submissions are mapped to reference

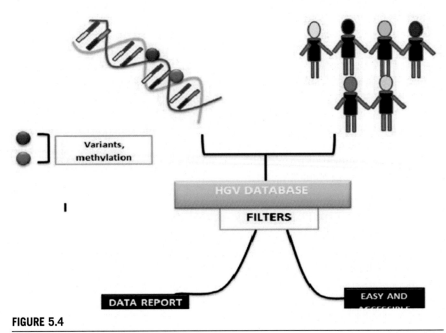

FIGURE 5.4

Workflow of HGV database.

sequences and presented. ClinVar then represents the information for interactive users and those who want to employ it for various purposes and applications. It collaborates with interested groups to address the genomics community's requirements as effectively and efficiently as feasible. ClinVar is intended to facilitate the advancement of our understanding of the connection between genotypes and clinically significant phenotypes. By putting together information about variants found in people with and without phenotypes, the database helps prove the clinical validity of human variation (Landrum et al., 2016).

dbSNP: NCBI and NHGRI (National Human Genome Research Institute) established and operate dbSNP, a publicly available resource documenting genetic variation within and between species. It comprises a variety of molecular variation, including SNPs, small insertion and deletion polymorphisms (indels/DIPs), short tandem repeats (STRs) or microsatellite markers, and heterozygous sequences (Sherry et al., 2001). Since its establishment in September 1998), dbSNP has functioned as a primary, publicly accessible database for genetic variation. Once these variants have been detected and cataloged in the database, additional laboratories will be able to access the sequencing data surrounding polymorphism and experimental settings in future research applications. dbSNP presently classifies nucleotide sequence variants according to the following categories and percentages:

I. single nucleotide substitutions, 99.77%;
II. short insertion/deletion polymorphisms, 0.21%;

III. invariant sequence sections, 0.02%;

IV. microsatellite repeats, 0.001%;

V. identified variants, 0.001%; and

VI. uncharacterized heterozygous tests, 0.001%.

The 1000 Genomes Project It was initiated in 2008 and seeks to create the most comprehensive atlas of human genetic diversity by sequencing around 2500 genomes from approximately 25 groups worldwide. The genetic variation data provided by this multinational partnership will aid genome-wide association studies of complex variables and phenotypes, such as the pharmacogenomic phenotype. The purpose of a GWAS is to identify genomic areas related to an outcome of interest (e.g., drug response). Frequently, the related variation acts as a marker (tag SNP) for the true causative variant (the variant that underlies the observed association). To identify the causative variant(s) and understand the mechanism behind its effects, more analyses of variations within the candidate area are necessary. In several instances, uncommon variations within the candidate locus contribute to disease susceptibility or phenotype. But these rare changes aren't in the HapMap database and aren't picked up by existing genotyping technologies, so they aren't taken into account in GWASs (Li et al., 2012).

5.4.2 Predicting the clinical significance of human genetic variation

Any type of testing, diagnosis, or risk assessment of the patient and their family for any type of genetic disease is called molecular genetic testing. Today, there has been a robust development in molecular genetic testing that is low-cost, fast, and efficient. The DNA sequencing technology and computational technologies combined increase the standard of molecular genetic testing, which shows the result that there is more DNA variants per test due to lifestyle and environmental factors, which may include pollution and global warming. To determine the pathogenicity of the DNA variants, various approaches and methods have been organized in a systemic way that can help overcome this increased number of DNA variants. Initially, the database provides existing data, which includes publication of internal and public resources. After that, the statistical analysis is being performed, which includes the assessment of population and disease. Now, both the in vivo and in vitro experiment data is being evaluated on the basis of computational prediction, which basically predicts the impact of each variant. Finally, with all the evidence, we can predict whether the variants will cause the disease or not (Duzkale et al., 2013).

A report suggests that the most common result of molecular genetic testing includes DNA variants and cystic fibrosis, or a gene that is linked with the disease and has a sequenced coding region. Next-generation technologies have improved diagnostic sensitivity while also providing high throughput results at a low cost. Such technology includes next-generation sequencing (NGS). It is also observed that the variants with uncertain significance (VUS) have also increased, which basically means that the genetic sequence for a particular disease is not clear. The functionality of the variant identified by the genetic testing is unclear. Identifying variants also helps us figure out what role a gene plays in a certain disease. Genome sequencing, on the other hand, adds layers of complexity to the questioning of the gene.

The molecular genetic testing has had great success in the diagnosis and evaluation of genetic diseases, providing high throughput results at a low cost. However, protocols, rules, regulations, and ethics are lacking in molecular genetic testing. When it comes to variant interpretation, there are some guidelines available that are updated as variant knowledge expands, including the American College of Medical Genetics and Genomics (ACMG). There are three questions that can be used to figure out what the clinical significance of a variant is for a certain disease.

a. Does genetic variation also change the functionality of a particular disease?
b. Is this variation cause disease?
c. Is there any relation between the associated disease and the present clinical condition of the patient?

While the interpretation of variant questions 1 and 2 is very important to assess the variant and its effect, But in some cases, to maximize the benefit to the patient, one should do the entire set of three questions, which provide a three-dimensional interpretation of the variant. This is especially true in the case of variants with uncertain significance (VUS). While testing 21,000–22,000 cases in a case study, it was discovered that 246 genes are linked to 50–56 human diseases. For clinical assessment, a systematic framework has been made, and more than 17,000 variants are being found and about 8000 of them are being recognized in patients.

The Fig. 5.5 shows the systematic representation of variant assessment. The genetic variation is acknowledged by the test which is performed by the labs. After that the results are being annotated on the basis of publication, internal database and computational prediction algorithm. The result which are got from annotation in the form of clinical data are being classified by the expert individual into five sections, benign, likely to benign, variants with uncertain significance (VUS), pathogenic, likely to pathogenic (McLaren et al., 2016).

5.5 Phenotype-genotype association

Genotype refers to the entire collection of genes present in an individual. The genotype is inherited from both parents and thus governs heredity and the extent of development. Sexually reproducing organisms have unique genotypes due to the fusion of gametes and the process of crossing over during meiosis. The characteristics of an individual are the phenotype and are governed by the genotype and other factors. Observable features or physical attributes of an individual or organism are referred to as their phenotype. Individual phenotypes are the result of a complex interaction between an individual's genotype and other factors such as epigenetic modifications

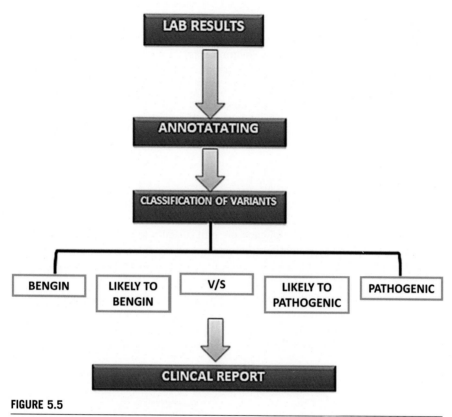

FIGURE 5.5

A typical variant assessment workflow.

and environmental factors. The phenotype includes color, height, shape, behavioral characteristics, biochemical processes, and many more. The phenotype is subjected to constant change due to ever-changing environmental conditions and the morphological and physiological changes associated with aging. The environmental factor influences the expression of the genotype. The association between genotype and phenotype is complex because, despite having 99.9% similar genomes, all individuals differ phenotypically to a great extent; Fig. 5.6 depicts the complex genotype-phenotype association.

The phenotype of an organism can be observed through simple, direct observation, whereas genotyping is done to observe or characterize the genotype. Genotyping involves various complex procedures, from the extraction of DNA to sequence analysis. Previously, genotyping of partial sequences was the only option, but with technological advancements, whole-genome sequencing can now be done quickly. Partial sequencing does not provide much information because most of the proteins are encoded by widely spread gene fragments, which may be present at different loci too. WGS makes it possible to identify the full sequence of a person, which can then

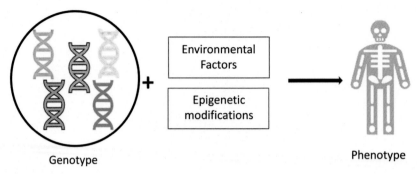

Genotype   Phenotype

FIGURE 5.6

Schematic representation of the complex relationship between genotype and phenotype.

be analyzed and compared to a phenotypic trait (usually to find the cause of a disease). Fig. 5.7 shows a schematic representation of the WGS workflow. The other methods used for genotyping include DNA microarrays, PCR, DNA hybridization, and others.

The understanding of genotype-phenotype association is a developing field of research, especially in the context of pharmacogenomics. Individual differences determine how an individual will react to a drug or how susceptible he or she is to contracting a disease. For example, variations in the CYC450A liver enzyme

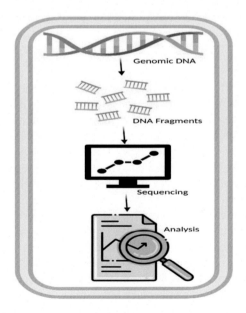

FIGURE 5.7

Schematic representation of the WGS workflow.

involved in the metabolism of several medicines, can influence a person's capacity for metabolizing a certain drug. This information about genetic differences can help doctors figure out the right amount of the drug to give for effective treatment.

Integrating both genotyping and phenotyping techniques is more adequate for a diagnosis or association study than genotype tests alone. The integrated approach is used in pharmacogenomics to identify the disease cause and drug metabolism with greater accuracy than predicted using genotyping alone. This integration also becomes the baseline for the personalized medicine approach. The genotype-phenotype association can easily be studied using knockout mice.

The field of genomics underwent a drastic transition with elucidating the human genome sequence and mapping of the SNPs (single nucleotide polymorphism). With technological advancement, it is now possible to analyze hundreds of SNPs at the same time. This feasibility in analysis has allowed the genotype-phenotype association to establish reproducibly. For example, SNPs of *TCF7L2* (transcription factor) and *PPARG* (peroxisome proliferator-activated receptor- γ) are associated with diabetes. Similarly, *NOD2* (nucleotide-binding oligomerization domain containing 2) SNP is associated to Crohn's disease, and *CFH* (complement factor H) SNP is associated with age-related macular degradation. Genotype-phenotype association studies are not only relevant for pharmacogenomics but are also essential to study the evolution of a particular trait. These studies help in the elucidation of genes responsible for behavioral traits. For most phenotypes, genes responsible are spread over different parts of the genome, which shape the phenotype differently under different environmental conditions. Due to this variable spread of genes encoding a phenotype, genome-wide associations are limited for such phenotypic traits.

Tremendous efforts have been made to link the human genome variations with the phenotype, and this process has accelerated over the past few decades. These studies have disclosed various phenotypes/traits affected by the variation present in the genome. To interpret the phenotypic prediction based on variation, a thorough understanding of genotype-phenotype association is required.

The various methods used for detection of genetic variation include the Mendelian approach of crossing and assessing the phenotype, but this method is not possible with humans due to the longer generation time. Other approaches include PCR based methods, FISH, RFLP, sequence analysis, and microarrays. These newer techniques are fast and precise, allowing multiple detections in a single pass.

5.6 **Pharmacogenomics**

Pharmacogenomics/pharmacogenetics refers to the study of genetic variations that cause variability in how different people react to drugs. In 1994, a study published in the USA presented the statistics that around 2.2 million people were severely affected by the drug side-effects, and around 100,000 people died due to the side-effects. According to these statistics, the death rate from drug side effects was higher than the death rate from viral diseases. Predicting how a patient will react to a specific drug will thus be a significant advance in the health sciences (Relling and Evans, 2015).

The individual's response to a particular drug depends on the target protein or enzyme and the receptors responsible for binding the drug and metabolizing it. Variations in such proteins at the genetic level may result in decreased binding or even the complete absence of drug binding to receptors and an absence of or decreased drug metabolism. The genetic composition of an individual can affect how the body reacts to the drug. Some factors that may be affected due to the genetic variation include:

5.6.1 Drug receptors

Certain drugs are attached to a particular target receptor to function effectively. The genetic composition determines the type, frequency, and specificity of the receptor, which may affect the way your body reacts to a particular drug. Depending upon the genotype an individual possesses, a higher or lower amount of drug may be needed, or a different drug may also be needed, as depicted in Fig. 5.8. Example: In the case of some breast cancers, *HER2* receptors are overexpressed, which helps in the development and progression of cancer. The *T-DM1* drug targets the *HER2* receptor of the cancerous cells and eventually kills them. If a person has breast cancer, then the clinician tests the tumor sample to check whether the samples are *HER2* positive or negative and then decides whether *T-DM1* can be used for treatment or not.

5.6.2 Drug uptake

Certain drugs need active uptake inside the target tissues or cells for their practical function. The genome of an individual affects the uptake of certain drugs. Decreased permeability toward a drug hinders the cellular uptake of the drug, leading to the accumulation of the drug at random sites, which may cause side effects. The genetic makeup of a person affects how efficiently a drug can be taken up by the target cells and how long it takes to be excreted out of the cells. If the drug is excreted too fast, it might not create the desired therapeutic effect, as shown in Fig. 5.9. Example: Statins are a class of drugs that help lower the cholesterol level by acting on the liver. For statin activity, they must be transported to the liver. This transportation is facilitated by a protein encoded by the gene SLCO1B1. A variation in the SLCO1B1 gene causes less uptake of a statin known as simvastatin into the liver. If administered at a higher dosage, simvastatin accumulates in the blood, which leads to problems in the muscles, which include pain and weakness. The clinician usually recommends the genetic testing of SLCO1B1 before prescribing simvastatin.

5.6.3 Drug breakdown

The genomic construction of an individual may affect the rate at which the drug is metabolized. If a drug gets metabolized faster than expected, it is excreted from the

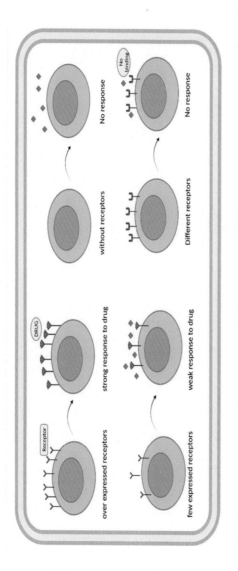

FIGURE 5.8

Variation in response to a particular drug due to genetic variations amongst individuals.

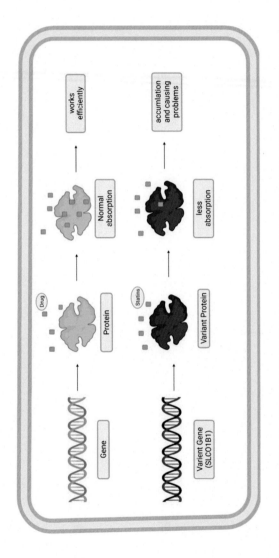

FIGURE 5.9

Variability in the drug uptake caused due to genetic variation.

body in a shorter period of time without causing the desired effect. In these cases, the drug dose needs to be increased or a different drug needs to be used. If the body metabolizes a drug more slowly than expected, a lower dosage is needed, as shown in Fig. 5.10. Example: The breakdown of amitriptyline, an antidepressant drug, is controlled by two genes named CYP2D6 and CYP2C19. A psychiatrist, before prescribing amitriptyline, generally recommends a genetic test for both genes. This testing helps identify the optimum dosage for the drug. If the genes are highly active, then the drug will breakdown faster, and either a high dosage will be prescribed, or a different drug will be prescribed. If the drug breakdown is slow, then a lower dosage will be prescribed.

One of the essential polymorphisms is found in the protein family, cytochrome P450. The enzyme CYP2D6 is essential for the breakdown/metabolism of about 20%–25% of all the drugs prescribed by physicians. Mutations in the gene encoding CYP2D6 can affect the rate of drug metabolism. Based on the type of mutation, the patients are classified as ultrafast drug metabolizers, extensive drug metabolizers, medium drug metabolizers, and slow drug metabolizers. This suggests that genetic polymorphism or genetic variation influences a patient's ability to respond to drugs. SNPs, being the most widely and frequently found genetic variation, are at the center of pharmacogenomics, as scientists are focused on determining the SNPs that alter the drug metabolism in an individual (Mateo et al., 2022).

Pharmacogenomics aims at predicting the unwanted side-effects of drugs used for therapy in advance. For predicting in advance, diagnostic test development is one of the prerequisites. This diagnostic test helps the clinician find out if any genetic predispositions are present in the patient and how the patient may react to a particular drug. Based on the diagnostic tests, the clinician or the geneticist finds out whether a distinct polymorphism is present in the drug-metabolizing enzymes or not. Based on the test results, the patients are then classified into various groups, and a suitable therapy is administered to them based on their genotype, as shown in Fig. 5.11. This type of therapy is also known as "stratified medicine" because the therapy is molded or tailored according to patients' genetic makeup.

Example of "Stratified Medicine": Mercaptopurine and thioguanine are used as chemotherapeutic agents to treat acute lymphatic leukemia (ALL). These drugs incorporate into the DNA of cancerous cells, eventually killing them. Thiopurine-S- methyltransferase is the enzyme responsible for the metabolism of these drugs. Genetic polymorphisms may interfere with the activity of this enzyme, leading to the accumulation of these drugs in blood cells and causing death. In contrast, in patients with high enzyme activity, high concentrations of drugs are needed. Pharmacogenomics also aids in the process of pharmacological research. Before entering the market, every drug must undergo rigorous clinical trials to prove its safety and efficacy. Pharmacogenomics helps exclude the cohort that may show no effect of the drug or may show unapprehensive side-effects before the start of the clinical trials. This filtering out of the cohort may increase the probability of the drug reaching the market. Pharmacogenomics also helps in the development of drugs for patients who do not respond to any of the therapies that already exist.

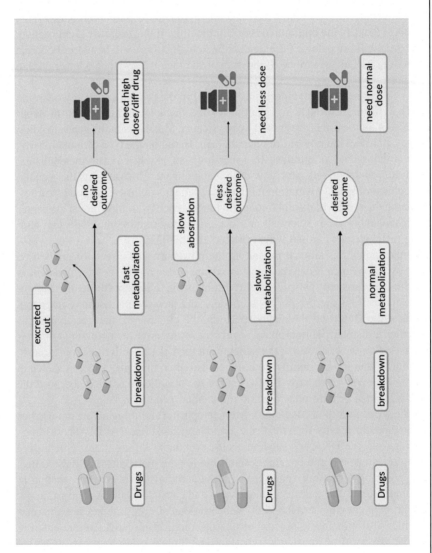

FIGURE 5.10

Schematic representation of different rates of drug breakdown.

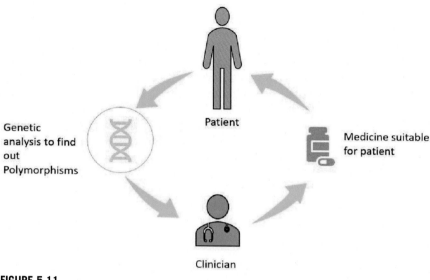

Patient

Genetic analysis to find out Polymorphisms

Medicine suitable for patient

Clinician

FIGURE 5.11

Schematic representation of finding genetic predispositions of the patient and tailoring drugs that would be more effective for treatment.

5.7 Pharmacogenomics and targeted drug development

Drug development using the pharmacogenomics approach focuses on treating the cause of the disease instead of just treating its symptoms. Some of the clinical diseases are caused by mutations in genes. Different types of mutations can be found in the same gene, which may have altogether different effects. Some mutations may cause a protein to lose its function, while other mutations may not lead to protein synthesis at all. Drugs can be designed based on the types of mutations that affect the protein. These drugs will be effective only against a specific type of mutation. Example: Ivacaftor is a drug used for the treatment of cystic fibrosis. A mutation in the CFTR gene leads to an alteration in the CFTR protein responsible for forming ion channels. The mutation causes this channel to be permanently closed, thus preventing the movement of particles. Ivacaftor forces the ion channel to open again, but this drug will be ineffective if the ion channels are not formed at all (Whirl-Carrillo et al., 2012).

Despite the genetic variation, other factors affecting the response and efficacy toward a drug include the patient's age, alcohol consumption, nutrition, the individual's microbiome, the presence of other diseases, and the consumption of other drugs. However, the presence of genetic variation can, but not necessarily will, lead to variation in metabolic pathways. So, to increase the efficiency of individual medicines, genetic predispositions and metabolic profiles should be considered. Thus, an integrated approach that incorporates both pharmacogenomics and pharmaco-metabolomics results in more tailored medicine.

5.7.1 Personalized medicine

Personalized medicine (PM) is a relatively new discipline of medicine that examines a person's genotype to aid in disease prevention, treatment, diagnosis, and cure. The genotype of a patient can be utilized to determine the optimal medicine or treatment option for them. The Human Genome Project's data is being used to enhance personalized treatment. Personalized medicine is a planned, effective, and preventative structural method for effective health care. To account for inherited predispositions, PM engages individuals in lifestyle decisions and effective management of health. PM is now possible and growing at a quicker rate thanks to significant technological advancements, including as: New methods for sequencing the human genome faster and more correctly with smaller but more powerful equipment (Whirl-Carrillo et al., 2012). Large-scale research and sample archives that link genetic variants to disease across countries and continents (Fig. 5.12).

5.7.2 Personalized medicine drivers

5.7.2.1 Human genome sequencing has been completed

The rise of information and knowledge outlining the causes of disease and patient heterogeneity, as well as heterogeneity in treatment response, will be the primary driver for customized medicine. The goal will always be to enhance the benefit-to- risk ratio associated with the administration of a particular medicine to a specific patient, as well as to identify newer, more specific, and safer medicines.

The emerging knowledge of medical science reveals that the heterogeneity of diseases and patients will almost certainly assist in the discovery of new disease-related objectives and more sensitive and accurate diagnostic tools. A standard, complex, chronic disease is thought to be caused by as many as 10 distinct genes on average. This should lead to a variety of new and improved therapeutic approaches as well as earlier interventions. As a result, healthcare practitioners will have a wide range of pharmacological options to treat patients with various diseases. When diagnostics packaged with pharmaceuticals become accessible, they will help health care providers decide which drug option is best for each patient. The health care provider will subsequently prescribe correctly. This decision will be based on drug efficacy, safety, or a combination of the two. Although no one knows when this will be possible, many, if not all, diseases will undoubtedly be several years away.

5.7.2.2 Molecular characterization of disease

As more information on the disease and patient heterogeneity becomes available, relevant genomic, genetic, and proteomic data linked to clinical information in well-characterized people will be collected and thoroughly mined utilizing bioinformatics and statistical methods. High-throughput genotyping, as well as gene and protein expression approaches, will be used to achieve this. This is already being done in several diseases, including cancer. It identifies somatic changes in gene expression and links them to interpatient variability in treatment response, as well as selecting appropriate therapeutic strategies. As a result of the timing of the

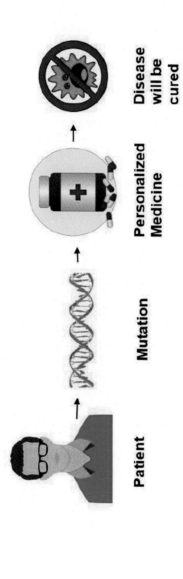

FIGURE 5.12

Schematic representation of the workflow of Personalized Medicine.

intervention, this could lead to a plethora of targets, and, eventually, medicines that will provide the most significant benefit to the illness classification will help to stratify at-risk patient populations further, resulting in early disease detection or prevention, more targeted therapeutics, and markers that can help guide drug therapy decisions. In some disorders, it may soon be possible to identify the genes and targets that have a significant role in disease development or penetrance.

Identifying all genes and targets involved in disease penetrance will take several years. However, using SNP maps, it is now possible to compare genetic differences between drug responders and non-responders. This could have an impact on how clinical trials are designed and on drug development for specific disorders. Similarly, while there is an increasing desire for safer pharmaceuticals, particularly in the treatment of chronic disease, there will be an increasing demand to use drugs in those who are most likely to benefit from them rather than those who will simply experience adverse effects. All of this will require the development of a number of markers and safety correlations that will help find the molecular cause of disease in a specific person.

5.7.2.3 Search for biomarkers of drug response

The role of environment, the level and complexity of the disease at a given time, the influence of drug-drug interactions, the person's overall health, including primary organ function, and disease complications are the parameters that lead to heterogeneity. In 1962, the first comprehensive treatment of pharmacogenetics was published. Because even the most acceptable medicines do not lead to substantial efficacy or safety in 100% of treated individuals, one option for personalized medicine could be to increase efficacy, safety, or both for an approved treatment. SAEs have been responsible for the withdrawal of drug candidates during drug development, as well as the removal of authorized pharmaceuticals after launch, due to genetic variance in drug efficacy. The role of environment, the level and complexity of the disease at a given time, the influence of drug-drug interactions, the person's overall health, including primary organ function, and disease complications are the parameters that lead to heterogeneity. In 1962, the first comprehensive treatment of pharmacogenetics was published. Because even the most acceptable medicines do not lead to substantial efficacy or safety in 100% of treated individuals, one option for personalized medicine could be to increase efficacy, safety, or both for an approved treatment. SAEs have been responsible for the withdrawal of drug candidates during drug development, as well as the removal of authorized pharmaceuticals after launch, due to genetic variance in drug efficacy (Fig. 5.13).

Biomarker-associated treatments for a variety of infections are obtainable and may explain clinical variability, drug response risks, particular doses, dose approaches, and polymorphic drug targets. The FDA-approved drugs for certain diseases along with their biomarkers are mentioned in Table 5.1. It describes medication that must be recommended to individuals with biomarker-related adverse effects. This can be extremely helpful in providing a safe and efficient

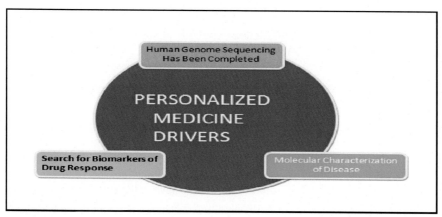

FIGURE 5.13

Factors contributing to the developments in the field of Personalized Medicines.

Table 5.1 Biomarkers of pharmacogenomics for the disease therapy.

Name of drug	Area of therapeutics	Biomarkers	Referenced subgroups	Efficacy
Afatinib (or tyrosine kinase inhibitor)	Oncology	EGFR	EGFR exon 19 deletion or exon 21 substitution (L858R) positive	Such changes bring sympathy in the treatment of afatinib
Carvedilol	Cardiology	CYP2D6	CYP2D6 poor metabolizers	Higher concentration of carvedilol plasma, dosage caution is needed
Celecoxib	Rheumatology	CYP2C9	CYP2D6 poor metabolizers	Lower dosage recommended
Diazepam	Psychiatry	CYP2C19	Poor metabolizers of CYP2C19	Lower dosage to rule out extended sedation
Abacavir	Infectious diseases (like HIV)	HLA-B	Allele carriers of HLAB5701	Higher risk of immune mediated hypersensitivity reactions

treatment based on biomarkers. The finding of novel and potential biomarkers associated with the infection may play a prominent role in the identification, placement, and behavior of the treatment itself.

5.7.3 Future aspects of pharmacogenomics in personalized medicine

Managing the severity of the condition is one of the most difficult aspects of treatment. In the following generation, a considerable number of novel mutations are observed. Although the relationship between some of these mutations and diseases is widely understood, the effect of multiple alterations/mutations has yet to be investigated. In disease forecasting, some changes may have individual or collective implications. Systematic retrieval, retrieval, and analysis are required to keep clinical data and make decisions. Better disease biomarkers can aid in diagnosis and treatment. Because a better diagnosis of the problem can improve the quality of treatment, there needs to be financial motivation to come up with new diagnoses.

Microbiome research is now possible thanks to advances in DNA sequencing. Microbiome research focuses on diseases and could lead to more personalized treatments. It is vital to identify the connection between genetic variation and its associated consequences.

Pharmacogenomics was born out of these organizations, and it no longer focuses just on clinical validation and application. The clinical utility and cost-effectiveness of pharmacogenetics tests should be considered. Data and trials relating to pharmacogenomics are only available for other medications, but this information isn't being turned into medical services on a large scale. However, the clinical use of pharmacogenomics testing is limited due to the disparity between its clinical performance and its use in individuals by health care professionals. Some drugs include pharmacogenomics-based CPIC dosing guidelines. Some strategic recommendations have been proposed because primary care practitioners lack adequate criteria to assure the success of tailored drugs. Some of the suggestions are to use more biomarkers, make a genetic code that protects privacy and anonymity, make sure drugs are available to everyone in a fair way, test and document genetic information, make sure genetic testing is regulated well, and make more people aware of drug-modified pharmaceuticals.

5.8 Computational biology methods for decision support in personalized medicine

Understanding studies of various levels and periods, like the incorporation of genetic information (i.e., genotypes) with a patient's data on medical history, remains an issue (i.e., phenotypes). Bioinformatics plays a very significant role in assisting integrative analyses at different levels to draw conclusions in support of more errorless detection & its solutions, as well as creating strict patient recommendations with desirable precautionary therapeutic options. Such strategies would enable health providers to provide their patients with the "proper information at the right time, in the correct form, to the correct people."

Data integration combines laboratory and clinical environments for translational bioinformatics methodologies, enabling more efficient management and cost reductions. To increase peculiar patient's medicine in the medical field, information at many levels, like cells, genes, proteins, medications, and functional aspects, must be incorporated within the phenotypic data in EHRs. Also, data incorporation will lead to the development of large databases to generate new information and skill proficiency. E.g., a web-based database with data mining and inquiry tools known as the RNA-SEQ Atlas was created for gene expression profiling. Other practical databases, microarray profiles, signaling pathways, & genetic ontologies are all linked through the integrative system.

Such comprehensive methods can be beneficial for comparing tissue-specific expression profiles to uncover trends, as well as correlating tissue functions with genetic modifications to enable other decision-making processes at multiple system levels. Collecting various origins of data that meet the domain & necessary inspection is a crucial stage in data integration. Many of these records are disorganized and include errors, and they must be repaired, cleaned, organized, and updated. Both redundancies and inconsistencies need to be eliminated and addressed, respectively. One gene, for example, could have numerous entries with various names. To bring the shared values together, such concerns must be handled using GO (gene ontology). Other techniques, such as the CABIG (Cancer Biomedical Informatics Grid), aid in the resolution of such complex problems.

Standardization is a critical step in resolving interoperability issues in several medical and laboratory settings across a wide range of knowledge domains. For example, the SNOMED CT (Systematized Nomenclature of Medicine Clinical Terms) & ICD (International Classification of Diseases) is frequently used for clinical data and billing processes. Clinicians use the Digital Imaging and Communications in Medicine (DICOM) standards to process imaging data. Rx Norm Resource is a database of drug terminologies that connects them. KD (knowledge discovery), a critical element of decision support, can be performed through a reiterative and dependent process involving data integration and data mining technologies.

These approaches are required for finding successful pharmacological targets and employing personalized treatments. For systematic studies and vigorous assessments, like chronobiology research, data mining is essential. Data mining techniques can identify spatiotemporal patterns and develop grouping, association, and dependency models. Neural networks, based on probability graph models like Bayesian networks, text data mining, and evolutionary design are just a few of the data mining techniques available. Agent-based modeling may describe complex nonlinear systems at various different living dimensions, from biological cells to different societies. For instance, many knowledge discovery and computational biology mythologies, such as BLAST, MATLAB, and clustering algorithms, are adopted.

Gene expression in lung tissue can be measured by these methods by inspecting the patterns of the dark/light circadian cycle. Dynamic thermal analysis (DTA) was discovered to be very helpful for the recognition and identification of cancer (breast cancer) using artificial neural networks (ANN). For inspecting the etiology of breast cancer to designate the multifaceted essence of the disease, an agent-form archetype of mammary ductal epithelium dynamics was utilized. It depicts the multifaceted essence of various diseases, which includes both cellular and molecular causes. Furthermore, semantic Web technologies have been recommended as helping organize and display pharmacogenomics knowledge related to the generation of various drugs and decision-making in various medical fields. Such informatics technologies have the potential to become critical components of CDSSs.

5.8.1 Pharmacogenomics information

In both personalized medical care and the pharmacogenetics field, databases are of prime importance. The main goal of the field of pharmacogenomics is to find biomarkers that can predict how toxic a drug will be and how the body will react to it. This will make patient treatment more personalized and effective.

Some of the pharmacogenomics Web Resources are:

1. PharmGKB (https://www.pharmgkb.org/)
2. CPIC (https://cpicpgx.org/)
3. DrugBank (https://go.drugbank.com/)
4. SCAN (http://www.scandb.org/)
5. PACdb (http://www.pacdb.org/)
6. Human Cytochrome P450 Allele Nomenclature database transitioned to PharmVar
7. Cytochrome P450 Drug Interaction Table (https://drugninteractions.medicine.iu.edu/MainTable.aspx)
8. FDA's pharmacogenetic website (https://www.fda.gov/drugs/scienceresearch/researchareas/pharmacogenetics/ucm083378.html)

5.8.1.1 Pharmacogenomics knowledgebase (PharmGKB)

The Pharmacogenomics Knowledgebase (Pharm GKB) contains genetic variants, explication, pharmacological routes, and drug action linkages. The effort is led by Stanford University and supported by the NIGMS (National Institutes of General Medical Sciences). PharmGKB has evolved into a crucial infrastructure for direct treatment over the last 20 years. Pharm GKB seeks to assist researchers in better understanding how genetic differences in people might influence medication responses. Pharm GKB seeks to assist researchers in better understanding how genetic differences in people might influence medication responses. PharmGKB assesses measurement recommendations by releasing dual genetic quotation guidelines and tying them to basic evidence on genetic and related medicines. A drug's name, generic name, short phrase, full description, including label quotes, and, in the case of US labels, whether the label is listed in the FDA Table for Pharmacogenetic Organizations are all used to define the label. PharmGKB collects data from a variety of sources, but it is up to the user to choose which data to utilize and how to administer the drug. The user is responsible for reviewing and drawing conclusions from all accessible guidelines. The FAIR principles are: available, useable, and active, and PharmGKB is dedicated to following them. These guidelines ensure that PharmGKB data is utilized in a way that may be accessed by others in the present and future. From top genomic and medication services, we give the names, chemical compositions, allele waves, geographical information, and variants (e.g., NCBI, PubChem, Ensemble, HGNC, dbSNP, gnomAD, etc.).

5.8.1.2 DrugBank

DrugBank is a comprehensive, publicly accessible digital resource that includes specific drug information, administration, action, and contact information for FDA-approved pharmaceuticals, including testing drugs produced through the FDA-approved method. DrugBank has become one of the most extensively used drug sources in the world due to its extensive, high-quality information. The general public, teachers, pharmacists, medical pharmacists, pharmacists, and the pharmaceutical business all use it regularly. The 2008 release of DrugBank 2.0 covered pharmacological, medical, and chemical information. DrugBank 3.0, which came out in 2010, improves medication-drug interactions, drug delivery information, and pharmacokinetic data. When DrugBank 4.0 came out in 2014, it had a lot of QSAR (work-related activity size) data, information about how drugs work, and ADMET statistics.

DrugBank's purpose is to give a comprehensive collection on pharmaceuticals, including-biochemical data, mechanisms, and targets, Pharmacological data. It concentrates on responses of drug related to changes in the gene of humans or particular polymorphisms.

DrugBank 5.0 has introduced a number of new data types. This contains information about the key drugs' effects on metabolite levels (pharmacometabolomics), genetic levels (pharmaco transcriptomics), and protein synthesis levels (pharmacoproteomics).

Over the years, we've collected and evaluated 27,572 peer-reviewed papers. The DrugBank blocking team has deleted, checked, and installed data from those sources that meet the entrance requirements. Many researchers have used this type of information to test new medication properties or recover old ones. The number of phase I, II, and III research pharmaceuticals in PathWhiz, a JavaScript-based picture rendering application, is used to display all of these ways. The DrugBank prevention team is also stepping up its efforts to collect more information on drugs, drug usage, and drug trafficking. Understanding drug pharmacokinetics, availability, and ADMET symptoms requires this knowledge. Drug linkages with Drug-Bank have been one of the most significant changes or improvements to current Drug-Bank 5.0 data. Details about drug interactions are critical for patients, doctors, and pharmacists, particularly when given to the elderly, and older patients.

5.8.1.3 CPIC

CPIC was formed in 2009 by collaborating with the pharmacogenomics research network (PRN) and the Pharm GKB. The goal of the CPIC is to offer specific gene/drug clinical practice recommendations to help integrate pharmacogenetic research into clinical practice. The CPIC gathers scientific information at all levels, from biological findings into clinical trials, then analyses and incorporates it into the standards. These CPIC recommendations will assist physicians in recognizing how a genetic test can be utilized to enhance medication therapy rather than outlining the reasons for testing. The CPIC currently comprises 174 drug/gene combinations from all CPIC levels, containing 63 genes and 132 therapeutics. Each CPIC recommendation has been reviewed by a group of experts and is kept up-to-date and available to the public on the CPIC website.

References

Duzkale, H., Shen, J., Mclaughlin, H., Alfares, A., Kelly, M., Pugh, T., Funke, B., Rehm, H., Lebo, M., 2013. A systematic approach to assessing the clinical significance of genetic variants. Clin. Genet. 84, 453−463. https://doi.org/10.1111/CGE.12257.

Eichler, E.E., 2019. Genetic variation, comparative genomics, and the diagnosis of disease. N. Engl. J. Med. 381, 64. https://doi.org/10.1056/NEJMRA1809315.

Fredman, D., Siegfried, M., Yuan, Y.P., Bork, P., Lehväslaiho, H., Brookes, A.J., 2002. HGVbase: a human sequence variation database emphasizing data quality and a broad spectrum of data sources. Nucleic Acids Res. 30, 387−391. https://doi.org/10.1093/NAR/30.1.387.

Landrum, M.J., Lee, J.M., Benson, M., Brown, G., Chao, C., Chitipiralla, S., Gu, B., Hart, J., Hoffman, D., Hoover, J., Jang, W., Katz, K., Ovetsky, M., Riley, G., Sethi, A., Tully, R., Villamarin-Salomon, R., Rubinstein, W., Maglott, D.R., 2016. ClinVar: public archive of interpretations of clinically relevant variants. Nucleic Acids Res. 44, D862−D868. https://doi.org/10.1093/NAR/GKV1222.

Lek, M., Karczewski, K.J., Minikel, E.V., et al., 2016. Analysis of protein-coding genetic variation in 60,706 humans. Nature 536, 7616. https://doi.org/10.1038/nature19057, 536, 285−291.

Li, M.J., Wang, P., Liu, X., Lim, E.L., Wang, Z., Yeager, M., Wong, M.P., Sham, P.C., Chanock, S.J., Wang, J., 2012. GWASdb: a database for human genetic variants identified by genome-wide association studies. Nucleic Acids Res. 40, D1047−D1054. https://doi.org/10.1093/NAR/GKR1182.

Mateo, J., Steuten, L., Aftimos, P., André, F., Davies, M., Garralda, E., Geissler, J., Husereau, D., Martinez-Lopez, I., Normanno, N., Reis-Filho, J.S., Stefani, S., Thomas, D.M., Westphalen, C.B., Voest, E., 2022. Delivering precision oncology to patients with cancer. Nat. Med. 28, 658−665. https://doi.org/10.1038/S41591-022-01717-2.

McLaren, W., Gil, L., Hunt, S.E., Riat, H.S., Ritchie, G.R.S., Thormann, A., Flicek, P., Cunningham, F., 2016. The ensembl variant effect predictor. Genome Biol. 17, 1−14. https://doi.org/10.1186/S13059-016-0974-4/TABLES/8.

Nei, M., 1983. Genetic polymorphism and the role of mutation in evolution. Evol. Genes Prot. 71, 165−190.

Relling, M.v., Evans, W.E., 2015. Pharmacogenomics in the clinic. Nature 526, 7573. https://doi.org/10.1038/nature15817, 526, 343–350.

Sherry, S.T., Ward, M.H., Kholodov, M., Baker, J., Phan, L., Smigielski, E.M., Sirotkin, K., 2001. dbSNP: the NCBI database of genetic variation. Nucleic Acids Res. 29, 308–311. https://doi.org/10.1093/NAR/29.1.308.

Studer, R.A., Dessailly, B.H., Orengo, C.A., 2013. Residue mutations and their impact on protein structure and function: detecting beneficial and pathogenic changes. Biochem. J. 449, 581–594. https://doi.org/10.1042/BJ20121221.

Whirl-Carrillo, M., McDonagh, E.M., Hebert, J.M., Gong, L., Sangkuhl, K., Thorn, C.F., Altman, R.B., Klein, T.E., 2012. Pharmacogenomics knowledge for personalized medicine. Clin. Pharmacol. Ther. 92, 414–417. https://doi.org/10.1038/CLPT.2012.96.

Structural bioinformatics

6

6.1 Introduction

The study of biology is done with the help of computer science and programming in the discipline of bioinformatics, which is an example of an interdisciplinary field. The advancement of biological research results in the collection of enormous volumes of biological data, which is difficult to process if one does not have access to computational capacity. When it comes to manipulating biological components, bioinformatics has proven to be a very useful tool. A subfield of bioinformatics known as structural bioinformatics investigates the structure of biological macromolecules and micromolecules, including things like protein, DNA, and RNA, among others. A living object is composed of a significant number of macromolecules. Therefore, in order to analyze them, you need to understand how they are constructed. The study of structural bioinformatics reveals to us how these components are assembled, so it assists us in learning more about them. Bioinformatics is an essential interdisciplinary topic that helps with storing, processing, and analyzing data in a methodical and comprehensive fashion. This is because there is a lot of biological data that is being made public. The use of bioinformatics approaches has made it simpler to manipulate biological data such as DNA sequences and protein sequences, thanks to the contributions of statistics and algorithms to the field. When compared to sequential data, biological structures are often more fluid and unpredictable. Because of this, the field of structural biology presents a more significant challenge than the field of bioinformatics.

Another part of structural bioinformatics is predicting the structure of biological macromolecules. Bioinformatics tools and algorithms can be used to figure out the structure of a protein. Proteins are the body's essential structural and functional components. Proteins are used to make hormones, which control how metabolism works. Proteins are also used to make hair, muscle fibers, antibodies, and wool, among other things. Enzymes are the most important part of all biochemical reactions that happen inside the body. They also help our bodies fight off infections, turn food into energy, and help with the replication, transcription, and translation of DNA. In general, about 60% of a person's body is made up of water, and 17% is made up of proteins. To do their jobs, proteins need to be folded into a certain shape and structure. By breaking down how proteins fold, researchers can better understand what they do in the body and might be able to make protein supplements for people who don't get enough. Misfolded peptides can give us more information about how diseases spread. A lot of work is being done to come up with new ways to test how proteins look in three dimensions. Researchers often use X-ray crystallography and NMR spectroscopy to deduce the structure of proteins. But the experimental methods cost a lot of money and take a long time. A protein's structure can be predicted based on its sequence or a similar structure.

Structural bioinformatics usually consists of the following steps:

- Collecting of biological data, either high throughput sequencing data or imaging data.
- Building a computational model can be a structural simulation, optimization, or alignment.
- Interpreting the model results from structural biology perspectives.
- Providing insights for the next iteration of experimental design.

6.2 Viewing protein structures

Visualization drives a cycle of experimentation, reasoning, speculation, and validation in many areas of research, particularly structural biology and biophysics. The use of molecular visualization in particular is now common in a variety of situations, whether it is to illustrate scientific research articles or to gain understanding of original research data. These techniques have long attracted widespread interest and demand. Data transfer from fundamentally three-dimensional objects to a 2D framework, such as paper or standard computer displays, is a significant challenge. The computer graphics discipline has made significant contributions to computer science. Because the field of structural biology requires easily accessible, end-user focused software tools for effective dissemination, these contributions could only arrive gradually. The long-standing need for molecular graphics has given rise to macromolecular structure visualization tools, which are now available to scientists of all backgrounds.

Molecular vaporization has a wide range of applications, from extremely general needs to specialized ones. A first generic application, for example, is the evaluation of hypotheses and the representation of scientific papers, such as when correlating mutational data with structural representations. The next level of application could be a more detailed visual examination of macromolecular structures and their attributes, potentially in relation to the (spatial) distribution of charges, electrostatic properties, pockets, and surface complementarities. An even more specialized application could be the representation and analysis of data using theoretical chemistry, computational biology, and bioinformatics methods. One notable example that has advanced the field is the need to examine increasingly complex molecular dynamics simulations. This application invariably leads to data analytics, specifically visual analytics.

Several molecular visualization software programmes have been in use by the community for many years and provide powerful visualization features to a large user base. Popular molecular visualization softwares include Chimera (the most recent version is called ChimeraX) (Pettersen et al., 2004), JMol (and more recently JSmol) (Herráez, 2006), PyMol (Yuan et al., 2017), and VMD (Humphrey et al., 1996). These legacy packages appear to be a safe choice for visualization-based applications, with a solid codebase and an accessible API. Although not always intuitive for beginners, their utility has grown over time, and they are well suited to molecular visualization tasks. Each tool may have a variety of features and provide extensions, scripts, and tutorials to improve the user's experience. Table 6.1 provides a brief overview of this indispensable set of software tools, while Fig. 6.1 depicts a visual comparison of a typical workday and the user interfaces of the various programmes. The Chimera package, originally known as Midas in 1976, provides features for visual exploration and the study of molecular structures. It is possible to examine related descriptive data, particularly cryo-EM datasets with density maps, molecular dynamics trajectories, and other data. It is easy to make animations of the systems shown, and there are many different ways to show them visually.

Jmol was founded in 1999. It is a versatile platform that can be used as a standalone viewer as well as in a web environment. Jmol is often used to show structures in educational applications because it is flexible and easy to add to courseware or use as a graphical interface for looking into structural databases.

PyMol, as the name implies, was released near the end of 1999 and is based on the Python scripting language. This tool is especially popular among experimentalists due to the properties it provides that are useful for crystallographic and NMR-derived structures. It generates illustrations suitable for publication and includes useful molecular editing and atom selection tools. The term VMD, or Visual Molecular Dynamics, refers to the software's primary focus since its inception in 1993, which has been the graphical interpretation of molecular dynamics data. It efficiently maintains even large systems and provides a wide range of enhancements, particularly for sophisticated visual analysis, via plugins.

All of these programmes have the ability to automate and script operations for easy re-use and bulk deployment, which is a critical feature. Despite their long

Table 6.1 Summary of four commonly used protein visualization tools.

Features/ Tools	VMD	Pymol	Chimera	JMol/JsMol
Scripting	TCL	Python	Python	Javascript
OS	Windows/Mac/Linux	Windows/ Mac/Linux	Windows/Mac/ Linux	Web, Windows/ Marc/Linux
URL	https://www.ks.uiuc. edu/Research/vmd/	https:// pymol.org/2/	https://www.cgl. ucsf.edu/ chimera/	http://jmol. sourceforge. net/

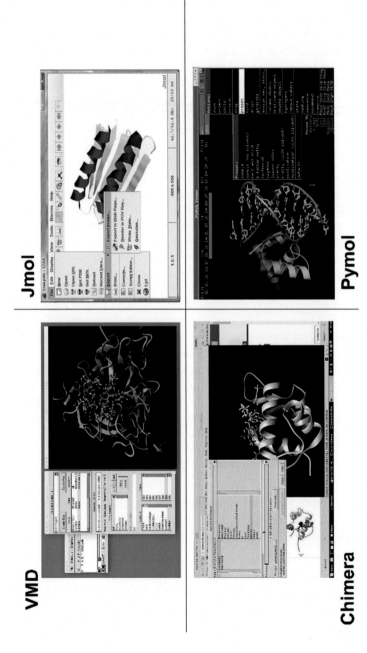

FIG. 6.1

Various protein visualization tool.

histories, all of the packages continue to add new features and adapt to changing hardware innovations, such as those in the graphics card market. Sometimes, as with ChimeraX, these additions necessitate a significant redesign of the underlying code.

6.3 Alignment of protein structures

The approach that is based on templates is the most trustworthy method for predicting the structure of macromolecules. At the moment, there are three distinct varieties of alignment-based strategies: those that are based on sequence, those that are based on profiles, and those that are based on structures. In most cases, the sensitivity of the profile-based alignment method is much higher than that of the sequence-based alignment systems. The structure-based alignment technique, on the other hand, is more sensitive than the profile-based alignment system. When it comes to matching, sequence-based tactics provide a higher level of specificity compared to profile-based strategies. In the same way that structure-based alignment methods are more particular, profile-based alignment tactics are as well.

When two proteins are identical, the sequence-based techniques will provide a better alignment than the other options. There are now a number of different techniques being developed, and the primary distinction between them is the operation of their alignment, gap penalty, and mutation score of amino-alkanoiate algorithms. Some alignment techniques, such as the Needleman-Wunsch algorithm for global alignment and the Smith-Waterman algorithm for local alignment, create the alignments via the use of dynamic programming. Other alignment strategies, such as the Smith-Waterman algorithm, generate the alignments locally. FASTA and BLAST are two examples of additional approaches that make use of alignment algorithms that are heuristically based. In order to determine the degree of similarity between two aligned residues, PAM and BLOSUM compares the aminoalkanoic acid substitution matrices that are extensively employed. Therefore, it is possible that two proteins with just a passing resemblance in their sequence would need to have the same structure in order to be considered homologous. The alignment standard is going to be strengthened with the help of the operational sequence profile. Multiple sequence alignment, also known as MSA, was used in the creation of the sequence profile by utilizing homologous sequences. Not only does it reveal the aminoalkanoic acid sequence, but it also gives information about other chemical processes. PSI-BLAST (Jones and Swindells, 2002) can be used to find close homologs of the target macromolecule in a large set of sequence data, such as the non-redundant NCBI data, so that an MSA can be built from these homologous elements, and then a sequence profile can be created from the multiple sequence alignment. PSI-BLAST can also be used to find close homologs of the target macromolecule in a large set of sequence data. There are several approaches to aligning the main sequence with a sequence profile, as well as numerous approaches to aligning two sequence profiles with one another. The process of matching a primary sequence to an HMM profile requires the use of two separate tools called HMMER and SAM square. The programmes "DIALIGN" (Morgenstern, 1999) and "FFAS" (Jaroszewski et al., 2011) are two further examples of sequences in profile alignment software. FORTE, HHpred, and Sculptor are examples of software programmes that are capable of user profile-to-profile alignment. Systems that move from one step to the next or that go from one step to a profile need to function less poorly than these ones do. In order to create a profile for a particular sequence of peptides, BLAST-PSI will be used. A profile-dependent alignment procedure likewise makes use of the sequence illustration profile as a standard measurement. PSI-BLAST reveals that a sequence-based profile is either a position-specific scoring matrix (PSSM) or a position-specific frequency matrix (PSFM). These matrices are heavily utilized in a variety of tools, such as those used for discovering similarities, determining folds, and predicting the structural properties of peptides. The length of the peptide sequence is denoted by N, and each position-specific scoring matrix (PSSM) and position-specific frequency matrix (PSFM) has 20N dimensions. When a PSFM is performed, each column maintains a record of the frequency with which 20 amino acids appear at a certain location in the sequence. Therefore, each column in a PSSM has the potential to transform into one of 20 unique amino acids in the same place. A reliable sequence-based profile will incorporate the maximum amount of information that can be included within the MSA. A sequence profile's quality is dependent not only on the samples it contains but also on factors such as the number of iterations of PSI-BLAST and the E-value cutoff that is used for determining whether or not two peptides are identical. When converting the frequency of aminoalkanoic acid to mutation potential, it further requires a method that may include pseudo-counts of aminoalkanoic acid. Profile Hidden Andrei Markov A different approach is used when modeling a multiple sequence alignment of homologous peptide sequences. The Hidden Andrei Markov Model is superior to the PSFM and PSSM because it explicitly models gaps in addition to taking into account correlations between neighboring residues. PSFM and PSSM only take into account correlations between adjacent residues. When it comes to aligning macromolecules and locating distant similarities, this indicates that profile HMM is, on average, more sensitive than PSFM/PSFM. A profile HMM might be in one of three states: match, insert, or delete. This is the most significant aspect. The "match" status of an MSA column indicates the likelihood that residues will be permitted to remain in the column once the analysis has been completed. and the next.

There are numerous tools used for the alignment of protein structure like-MADOKA (Deng et al., 2019), iPBA (Gelly et al., 2011), protein tertiary structure, pairwise sequence alignment, multiple sequence alignment, etc. (Figs. 6.2 and 6.3).

6.4 Structural prediction

Through the use of sequence similarity searches, multiple sequence alignments, identifying and describing domains, predicting secondary structure, predicting

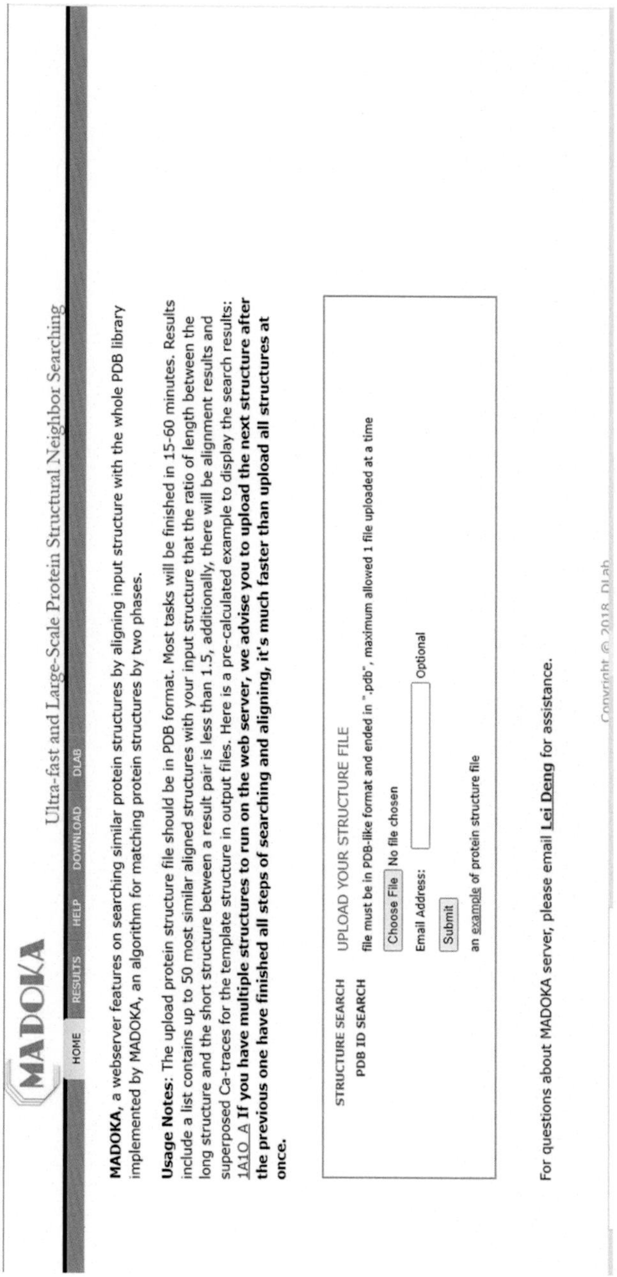

MADOKA
Ultra-fast and Large-Scale Protein Structural Neighbor Searching

HOME   RESULTS   HELP   DOWNLOAD   DLAB

MADOKA, a webserver features on searching similar protein structures by aligning input structure with the whole PDB library implemented by MADOKA, an algorithm for matching protein structures by two phases.

Usage Notes: The upload protein structure file should be in PDB format. Most tasks will be finished in 15-60 minutes. Results include a list contains up to 50 most similar aligned structures with your input structure that the ratio of length between the long structure and the short structure between a result pair is less than 1.5, additionally, there will be alignment results and superposed Ca-traces for the template structure in output files. Here is a pre-calculated example to display the search results: 1A10_A **If you have multiple structures to run on the web server, we advise you to upload the next structure after the previous one have finished all steps of searching and aligning, it's much faster than upload all structures at once.**

STRUCTURE SEARCH   UPLOAD YOUR STRUCTURE FILE
PDB ID SEARCH   file must be in PDB-like format and ended in ".pdb", maximum allowed 1 file uploaded at a time
                        Choose File No file chosen
                        Email Address: [   ] Optional
                        Submit
                        an example of protein structure file

For questions about MADOKA server, please email **Lei Deng** for assistance.

Copyright © 2018, DLab

FIG. 6.2

Image of MADOKA webpage.

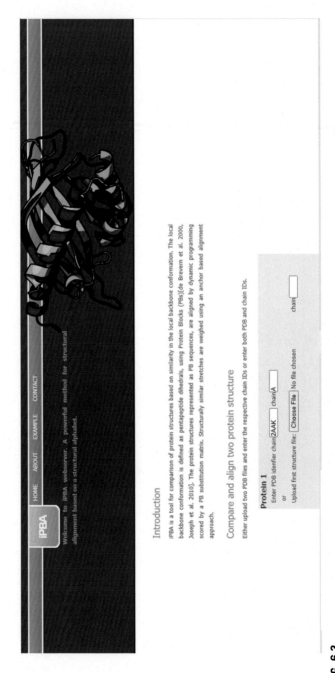

FIG. 6.3

Image of iPBA webpage.

solvent accessibility, automatically recognizing protein folds, building three-dimensional models down to the atomic level, and validating models, bioinformatics can be used to predict structures. Validating models is another application of bioinformatics. All of these approaches are not being used in every investigation into the composition of a protein's three-dimensional structure. Finding a reliable structural target from which to infer information about a protein's three-dimensional shape in response to a query sequence is an essential aspect of the process of protein structure prediction. There are three distinct types of schemes, each of which is determined by how it is carried out. The first step is to implement methods that are standard and well-known in the industry. In the event that a structural template cannot be found, the second effort has to make use of procedures that are more difficult. If a target fold cannot be defined in a reliable manner because many studies of nontrivial data have yielded varying findings, the design falls into the third category, and it will be very complicated, if not impossible, to complete with any degree of dependability.

Experiments that scientists do in order to determine the sequence and structure of proteins have seen a lot of innovation and improvement in the course of the previous few decades. Since the 1990s, the amount of protein data that has been uploaded to UniProt2 and the Protein Data Bank (PDB) has increased at a pace that is virtually exponential. When compared to obtaining the structures of proteins, obtaining their sequences is a much simpler task. The rapid accumulation of protein sequence data may be attributed to the advent of more sophisticated technologies for DNA sequencing. The UniProt/TrEMBL database now contains more than 85 million different protein sequences. X-ray crystallography and nuclear magnetic resonance (NMR) spectroscopy are the two primary experimental approaches that are used to find out how proteins are put together. Both of these methods focus on the structure of proteins. But both methods require a lot of time and work, and each has its own technical limitations when it comes to studying the proteins they want to study (Table 6.2).

6.4.1 Use of sequence patterns for protein structure prediction

The sequences of a genome include a wealth of information on the function of proteins and the ways in which they have evolved through time. This knowledge may be put to use to discover evolutionary connections between protein residues and to address the age-old challenge of determining the three-dimensional structure of a protein based on its amino acid sequence. Recent work on this issue has resulted in a significant amount of improvement, which can be attributed to the rapidly increasing number of sequences that are now accessible as well as the use of global statistical approaches. A better understanding of covariation, in addition to the three-dimensional structure, may help in the search for functional residues that are involved in the formation of protein complexes, the binding of ligands, and changes in conformation. The computer prediction of protein structures, which has been an issue in molecular biology for more than 40 years, may be able to fill this gap if it can be done with adequate precision. By comparing the amino acid sequence of interest

Table 6.2 Tools for protein prediction.

Name	Used for	Description
IntFOLD (Mcguffin et al., 2019)	Used for prediction of tertiary structure—3D modeling, domain prediction	The 3D structure and function may be predicted from amino acid sequences using an automated and integrated workflow.
RaptorX (Wang et al., 2016)	Is used for detection of remote homology	Progressed in predicting 3D structure whose protein sequence is without close homologous in PDB
Biskit (Grünberg et al., 2007)	3D modeling, template identification, and alignment	Predicting protein structure using homology modeling
ESyPred3D (Lambert et al., 2002)	Protein design and energy calculations	Using neural networks, the technique provides better alignment. Incorporating, weighting, and filtering various alignment processes provides alignments.
Rosetta (Baek et al., 2021)	Short fragments of known proteins are assembled by a Monte Carlo strategy to yield native-like protein conformation	De novo protein structure prediction
AlphaFold (Jumper et al., 2021)	Protein structure prediction	An artificial intelligence program developed by DeepMind, a subsidiary of alphabet, which performs predictions of protein structure.

to the sequence of another protein whose three-dimensional structure is known, several useful and fairly accurate three-dimensional models have been generated from amino acid sequences. These models may be used to better understand how proteins work. Building a model using a template, is often known as homology modeling. It has been proven difficult to produce good de novo predictions from the sequence when there is no known structure in a protein family. To fold proteins from scratch, some of the best, most current, and most cutting-edge methods, such as Rosetta, involve scanning for fragments with similar sequences in databases of three-dimensional structures and then utilizing empirical intermolecular force fields to put those fragments together. These kinds of approaches have been shown to be effective for proteins containing less than 90 amino acids, but in order to analyze bigger proteins, they need to be paired with the results with iterative tests of predicting structure of small residue patches. Other methods make an effort to generate 3D structure through predicting residue contacts by combining three-dimensional information with machine-learning techniques such as support vector machines, random forests, and neural networks. Despite these efforts, contact prediction accuracy remained "still quite low," and substantial improvements to models were only achieved for some small proteins. The issue of de novo structure prediction does not scale, meaning that the conformational search space grows exponentially as the size of the protein does. This presents a basic computational hurdle, even for approaches that are based on fragments of the protein. In this way, the main problem of predicting de novo structures in three dimensions has not been solved satisfactorily.

6.4.2 Prediction of protein secondary structure from the amino acid sequence

The local structure that is formed by a protein's polypeptide backbone is referred to as the protein's secondary structure. The -helix (H), the -strand (E), and the coil (C) area are the three different kinds of secondary structures. While states H and E both exhibit regular behavior, the C state does is disordered. The Dictionary of Secondary Structure of Proteins (DSSP) (Kabsch and Sander, 1983) was the product of Sander's efforts to develop a system for the assignment of secondary structure. The patterns of hydrogen bonds are used by this approach to automatically categorize the secondary structure into eight different states, which are labeled H, E, B, T, S, L, and G. Most of the time, these eight states may be further simplified into only three categories: helix, sheet, and coil. Usually, the helix is referred to by the letters G, H, and I; the sheet is referred to by the letters B and E, and all other states are referred to as a coil. The challenge of predicting the secondary structure of a protein appears as follows: given the sequence of amino acids that make up the protein, determine whether or not each amino acid is located in the α-helix (H), β-strand (E), or coil area (C). The prediction of the protein's secondary structure is often evaluated based on its Q3 accuracy. This metric, which evaluates the proportion of residues for three-state secondary structures, determines whether or not the structures have been successfully predicted.

In 1951, Pauling and Corey made the hypothesis that the backbones of protein polypeptides might assume helical or sheet-like forms. This was in the days before the first structure of a protein was discovered. A great number of statistical and machine learning approaches have been developed in order to determine the secondary structure. One of the first approaches to predicting the secondary structure of a protein combined statistical analysis with heuristic analysis. In order to overcome the issue of correctly forecasting the secondary structure, the GOR technique makes use of information theory. The position-specific scoring matrix (PSSM), which comes from PSI-BLAST, shows how proteins have changed over time and what has caused the biggest changes in how secondary structures of proteins are predicted. Protein composition is the consideration of a shape along with 3 protein components from an amino acid series—the prediction of the second and higher structure from the primary. The prediction of the structure is unique to the ever-changing hassle of protein formation. Guessing protein composition is one of the most vital dreams pursued by way of laptop biology; and is critical in remedy (for instance, in the manufacture of medicine) and biotechnology.

6.4.3 **Chou Fasman method**

The Chou-Fasman method (Chou and Fasman, 1978) is a computational method for predicting the secondary structure of proteins from their primary sequence. Protein secondary structure refers to the local, regular conformations adopted by the poly-peptide chain, which are formed by the interactions between the peptide bonds and the side chains of the amino acids. The main types of secondary structure are the alpha helix, beta sheet, and random coil. The Chou-Fasman method is based on the idea that the amino acid sequence of a protein can influence its secondary structure. Specifically, the method uses statistical analysis to identify patterns in the distribution of amino acids in known protein structures and then uses these patterns to predict the secondary structure of a protein from its primary sequence.

To perform the prediction, the protein sequence is first divided into overlapping windows, and the amino acid composition of each window is analyzed. The method then assigns a score to each amino acid based on its propensity to be found in different types of secondary structure. For example, amino acids with hydrophobic side chains, such as leucine and valine, tend to be found in alpha helices, while amino acids with polar side chains, such as serine and threonine, tend to be found in beta sheets. The method then uses these scores to calculate the likelihood that a particular amino acid will be found in a particular type of secondary structure. This likelihood is expressed as a probability, and the secondary structure with the highest probability is predicted for each amino acid. One of the advantages of the Chou-Fasman method is that it can be applied to proteins of any size and is relatively easy to implement. However, the method has some limitations. First, it relies on sta-tistical analysis and may not always be accurate. Second, it can only predict the secondary structure of a protein, not its tertiary structure, which refers to the three-dimensional shape of the protein as a whole. Finally, the method does not take into account the effects of the protein's environment, such as pH and tempera-ture, which can also influence its secondary structure. Despite these limitations, the Chou-Fasman method remains a widely used tool in protein structure prediction and has contributed to our understanding of the relationship between a protein's primary sequence and its secondary structure. It has also been used in combination with other methods, such as hydrogen bonding analysis and structural alignments, to improve the accuracy of protein structure prediction.

6.4.4 **GOR method**

The GOR (Gödel, Or, and Roth) (Garnier et al., 1978) method is a widely used method for predicting the secondary structure of proteins from their amino acid sequence. It is based on the idea that the local sequence of a protein is related to its local conformation, and that the conformation of a protein can be predicted by looking at the sequence of amino acids around it. The GOR method uses a statistical approach to predict the secondary structure of a protein based on the sequence of its amino acids. It does this by building a statistical model based on a set of known protein structures, and then using this model to make predictions about the secondary structure of a given protein. The GOR method begins by dividing the protein into overlapping segments, called windows, and then looking at the sequence of amino acids within each window. It then compares the sequence of each window to the sequences of windows in known protein structures, and uses this information to predict the secondary structure of the protein. The GOR method has a number of advantages. One of the main advantages is that it is relatively fast, as it only requires a single pass through the protein sequence. Additionally, the GOR method is relatively accurate, with an average prediction accuracy of around 70%.

There are also a number of limitations to the GOR method. One of the main limitations is that it only predicts the secondary structure of a protein, and does not provide any information about the tertiary structure or overall folding of the protein. Additionally, the GOR method is based on statistical models, and is therefore subject to the limitations of such models. Finally, the GOR method may not work well for proteins with unusual or atypical sequences, as it is based on a set of known protein structures and may not be able to accurately predict the structure of proteins that do not fit this set. Overall, the GOR method is a useful tool for predicting the secondary structure of proteins, and has played a significant role in the study of protein structure and function. It is particularly useful for predicting the secondary structure of proteins in the early stages of protein structure prediction, and can provide valuable information about the local conformation of a protein. However, it is important to be aware of its limitations, and to use it in conjunction with other methods in order to obtain a more complete picture of protein structure (Fig. 6.4).

6.4.5 **Prediction of three-dimensional protein structure**

Predicting the three-dimensional (3D) structure of a protein from its amino acid sequence is a central problem in computational biology. The 3D structure of a protein plays a critical role in its function, and determining the structure of a protein can provide insights into how it performs its function, how it may be modified or regulated, and how it may interact with other molecules. There are several experimental techniques that can be used to determine the 3D structure of a protein, such as X-ray crystallography and nuclear magnetic resonance (NMR) spectroscopy. However, these techniques are time-consuming and costly, and may not be feasible for all proteins. In addition, many proteins do not crystallize well or are too large to be studied by NMR, making it difficult to determine their 3D structure using these techniques.

Computational methods offer a faster and potentially more cost-effective alternative for predicting the 3D structure of a protein. These methods can be broadly classified into two categories: homology modeling and de novo prediction. Homology modeling involves using the 3D structure of a related protein, or "template," as a starting point to predict the 3D structure of a protein of interest. This approach relies on the assumption that proteins with similar amino acid sequences will have similar 3D structures. The protein of interest is first aligned with the template, and the resulting sequence alignment is used to build a model of the protein's 3D structure.

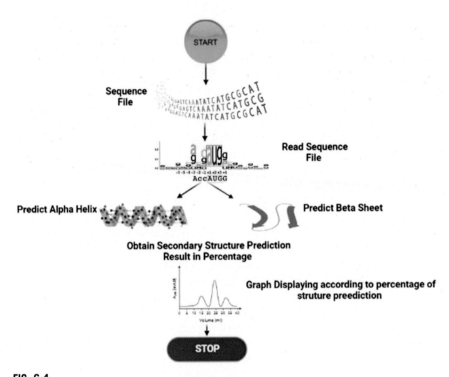

FIG. 6.4

GOR method.

Homology modeling can be an effective approach for predicting the 3D structure of a protein if a suitable template is available and the sequence identity between the protein of interest and the template is high.

De novo prediction, on the other hand, involves predicting the 3D structure of a protein from scratch without using a template. This approach relies on the physical and chemical properties of the amino acids and the interactions between them. De novo prediction methods can be divided into two main categories: physics-based methods and knowledge-based methods. Physics-based methods use physical principles, such as energy minimization, to predict the 3D structure of a protein. These methods are based on the idea that the protein will adopt a conformation that minimizes its energy. Knowledge-based methods, on the other hand, use statistical information about the 3D structures of known proteins to predict the 3D structure of a protein. These methods rely on the assumption that proteins with similar amino acid sequences will have similar 3D structures. Threading, also known as fold recognition, involves searching for the best fit of the target protein's amino acid sequence onto a library of known 3D protein structures. This approach is relatively fast and can be accurate, but it is limited by the size and quality of the library of known protein structures.

There are a number of different algorithms and software tools available for predicting the 3D structure of a protein using computational methods. Some of the most commonly used methods include ROSETTA, Modeler, and I-TASSER. These tools are widely used by researchers in academia and industry to predict the 3D structure of proteins and to understand their function.

Despite advances in computational methods, predicting the 3D structure of a protein remains a challenging problem. Accurate prediction of protein structure can provide valuable insights into the function of a protein and can be useful in drug design and other applications. However, current methods are often limited in their accuracy, and further research is needed to improve the prediction of protein structure. While these methods are not perfect and cannot always accurately predict the 3D structure of a protein, they offer a valuable alternative to experimental techniques and have the potential to significantly accelerate our understanding of protein structure and function (Fig. 6.5).

6.4.6 Evaluating the success of structure predictions

Evaluating the success of computational protein structure predictions is an important aspect of the protein structure prediction field, as it allows researchers to determine the accuracy and reliability of different prediction methods. There are several different ways to evaluate the success of protein structure predictions, and the most appropriate method will depend on the specific goals of the prediction and the type of data available. One common method for evaluating the success of protein structure predictions is to compare the predicted structure to the experimentally determined structure, using a metric known as the root mean square deviation (RMSD). The RMSD measures the average distance between the atoms in the predicted and experimental structures, and a lower RMSD indicates a more accurate

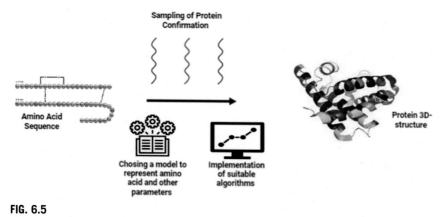

FIG. 6.5

3D protein structure prediction.

prediction. The RMSD can be calculated for all atoms in the protein, or for a subset of atoms such as the backbone atoms or side chain atoms.

Another metric that is often used to evaluate protein structure predictions is the precision and recall of the prediction. Precision measures the fraction of predicted residues that are correctly predicted, while recall measures the fraction of correctly predicted residues out of all residues in the protein. A high precision and recall indicates a more accurate prediction. Another way to evaluate protein structure predictions is to use a benchmarking dataset, which consists of a set of proteins for which the experimental structure is known. The prediction method is applied to each protein in the dataset, and the accuracy of the prediction is compared to the experimental structure. This allows researchers to compare the performance of different prediction methods on a standardized dataset. In addition to these quantitative measures, it is also important to consider the biological relevance of the predicted structure. A prediction that is highly accurate according to the RMSD or precision and recall metrics may not necessarily be biologically relevant if it does not correctly capture important features of the protein such as its active site or binding site. Overall, evaluating the success of computational protein structure predictions is a complex task that involves considering a range of different metrics and factors. By carefully considering the accuracy, precision, and biological relevance of the prediction, researchers can determine the effectiveness of different prediction methods and identify areas for improvement.

EVA (Eyrich et al., 2001) is a web server that evaluates the effectiveness of automated approaches for predicting the three-dimensional structure of proteins. The assessment is set to be automatically updated once per week to ensure that it remains current despite the ever-increasing number of prediction servers and the ever-shifting methods for making predictions. Every day, the servers connect to the Protein Data Bank (PDB) to get the sequences of newly discovered protein structures. Their predictions are then put together.

References

Baek, M., DiMaio, F., Anishchenko, I., Dauparas, J., Ovchinnikov, S., Lee, G.R., Wang, J., Cong, Q., Kinch, L.N., Dustin Schaeffer, R., Millán, C., Park, H., Adams, C., Glassman, C.R., DeGiovanni, A., Pereira, J.H., Rodrigues, A.v., van Dijk, A.A., Ebrecht, A.C., Opperman, D.J., Sagmeister, T., Buhlheller, C., Pavkov-Keller, T., Rathinaswamy, M.K., Dalwadi, U., Yip, C.K., Burke, J.E., Christopher Garcia, K., Grishin, N.v., Adams, P.D., Read, R.J., Baker, D., 2021. Accurate prediction of protein structures and interactions using a three-track neural network. Science 373, 871−876. https://doi.org/10.1126/SCIENCE.ABJ8754/SUPPL_FILE/ABJ8754_MDAR_REPRODUCIBILITY_CHECKLIST.PDF.

Chou, P.Y., Fasman, G.D., 1978. Prediction of the secondary structure of proteins from their amino acid sequence. Adv. Enzymol. Relat. Area Mol. Biol. 47, 45−148. https://doi.org/10.1002/9780470122921.CH2.

Deng, L., Zhong, G., Liu, C., Luo, J., Liu, H., 2019. MADOKA: an ultra-fast approach for large-scale protein structure similarity searching. BMC Bioinf. 20, 1–10. https://doi.org/10.1186/S12859-019-3235-1/FIGURES/5.

Eyrich, V.A., Martí-Renom, M.A., Przybylski, D., Madhusudhan, M.S., Fiser, A., Pazos, F., Valencia, A., Sali, A., Rost, B., 2001. EVA: continuous automatic evaluation of protein structure prediction servers. Bioinformatics 17, 1242–1243. https://doi.org/10.1093/BIO-INFORMATICS/17.12.1242.

Garnier, J., Osguthorpe, D.J., Robson, B., 1978. Analysis of the accuracy and implications of simple methods for predicting the secondary structure of globular proteins. J. Mol. Biol. 120, 97–120. https://doi.org/10.1016/0022-2836(78)90297-8.

Gelly, J.C., Joseph, A.P., Srinivasan, N., de Brevern, A.G., 2011. iPBA: a tool for protein structure comparison using sequence alignment strategies. Nucleic Acids Res. 39, W18. https://doi.org/10.1093/NAR/GKR333.

Grünberg, R., Nilges, M., Leckner, J., 2007. Biskit—a software platform for structural bioinformatics. Bioinformatics 23, 769–770. https://doi.org/10.1093/BIOINFORMAT-ICS/BTL655.

Herráez, A., 2006. Biomolecules in the computer: Jmol to the rescue. Biochem. Mol. Biol. Educ. 34, 255–261. https://doi.org/10.1002/BMB.2006.494034042644.

Humphrey, W., Dalke, A., Schulten, K., 1996. VMD: visual molecular dynamics. J. Mol. Graph. 14, 33–38. https://doi.org/10.1016/0263-7855(96)00018-5.

Jaroszewski, L., Li, Z., Cai, X.H., Weber, C., Godzik, A., 2011. FFAS server: novel features and applications. Nucleic Acids Res. 39, W38–W44. https://doi.org/10.1093/NAR/GKR441.

Jones, D.T., Swindells, M.B., 2002. Getting the most from PSI–BLAST. Trends Biochem. Sci. 27, 161–164. https://doi.org/10.1016/S0968-0004(01)02039-4.

Jumper, J., Evans, R., Pritzel, A., Green, T., Figurnov, M., Ronneberger, O., Tunyasuvunakool, K., Bates, R., Žídek, A., Potapenko, A., Bridgland, A., Meyer, C., Kohl, S.A.A., Ballard, A.J., Cowie, A., Romera-Paredes, B., Nikolov, S., Jain, R., Adler, J., Back, T., Petersen, S., Reiman, D., Clancy, E., Zielinski, M., Steinegger, M., Pacholska, M., Berghammer, T., Bodenstein, S., Silver, D., Vinyals, O., Senior, A.W., Kavukcuoglu, K., Kohli, P., Hassabis, D., 2021. Highly accurate protein structure prediction with AlphaFold. Nature 596, 7873. https://doi.org/10.1038/s41586-021-03819-2, 596, 583–589.

Kabsch, W., Sander, C., 1983. Dictionary of protein secondary structure: pattern recognition of hydrogen-bonded and geometrical features. Biopolymers 22, 2577–2637. https://doi.org/10.1002/BIP.360221211.

Lambert, C., Léonard, N., de Bolle, X., Depiereux, E., 2002. ESyPred3D: prediction of proteins 3D structures. Bioinformatics 18, 1250–1256. https://doi.org/10.1093/BIOINFOR-MATICS/18.9.1250.

Mcguffin, L.J., Adiyaman, R., Maghrabi, A.H.A., Shuid, A.N., Brackenridge, D.A., Nealon, J.O., Philomina, L.S., 2019. IntFOLD: an integrated web resource for high performance protein structure and function prediction. Nucleic Acids Res. 47, W408–W413. https://doi.org/10.1093/NAR/GKZ322.

Morgenstern, B., 1999. Dialign 2: improvement of the segment-to-segment approach to multiple sequence alignment. Bioinformatics 15, 211–218. https://doi.org/10.1093/BIOIN-FORMATICS/15.3.211.

Pettersen, E.F., Goddard, T.D., Huang, C.C., Couch, G.S., Greenblatt, D.M., Meng, E.C., Ferrin, T.E., 2004. UCSF Chimera—a visualization system for exploratory research and analysis. J. Comput. Chem. 25, 1605—1612. https://doi.org/10.1002/JCC.20084.

Wang, S., Li, W., Liu, S., Xu, J., 2016. RaptorX-Property: a web server for protein structure property prediction. Nucleic Acids Res. 44, W430. https://doi.org/10.1093/NAR/GKW306.

Yuan, S., Chan, H.C.S., Hu, Z., 2017. Using PyMOL as a platform for computational drug design. Wiley Interdiscip. Rev. Comput. Mol. Sci. 7, e1298. https://doi.org/10.1002/WCMS.1298.

High throughput technology

7.1 Omics theory

The flow of genetic information was described in late 1940s as unidirectional where DNA to RNA and RNA to Protein, However, it was also proposed that DNA comes from RNA and which is later translated to protein. With a lot of confusion in late 40s, however, in late 50s Francis crick has proposed that flow of information through gene to protein but it was found difficult to explain the flow of information from protein to protein. This information is determination of précis sequence of précis information of sequences. Afterward, there were number of different modifications by Francis crick and others postulated the correct genetic flow and residue by residue transfer of sequenced information although this information was quite intriguing but did not satisfy many (Crick, 1970). However other theory, regarding central dogma was also described by Watson but that was not valid in scientific fraternity. Finally crick's theory of Central dogma was accepted and genetic flow is DNA to RNA, so the process of transcription and RNA to Protein through translation. However, a modification has been done and that is production of DNA from RNA through process called Reverse transcription. In the Omics theory and central dogma of life can be described as (Fig. 7.1):

(1) DNA formation from DNA with the help of DNA polymerase, process is called Replication.
(2) Formation of RNA from DNA with the help of RNA polymerase, process is called transcription.
(3) Conversion of RNA to protein and this process is called translation in ribosome.
(4) Another process where RNA is reversely transcribed to form DNA through reverse transcription using reverse transcriptase.

7.2 High-throughput technologies

Automation equipment along with the classical methodology of cell biology in order to target biological questions that were not met using conventional methods. Techniques such as optics, biology, chemistry, and image analysis for much faster parallel studies of cell functioning, interaction, and diseases caused due to pathogenesis.

All About Bioinformatics. https://doi.org/10.1016/B978-0-443-15250-4.00003-4

OMICS THEORY

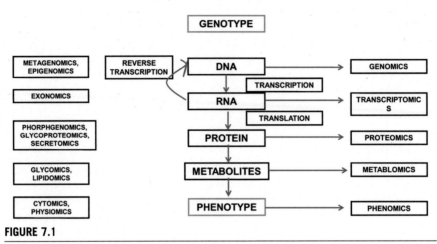

FIGURE 7.1

Flow chart: Omics theory.

High-throughput srceening includes a large number of samples of cells with model disease and various compounds extracted from specific sources using a computer for determining the compound of interest has the desired effect on the samples of cells (Szymański et al., 2012).

In discovering drugs, Sorafenib (Nexavar) used in medication therapy for treating different cancers, including renal cell carcinoma in the kidneys, thyroid cancer, and liver cancer. Sorafenib helps in order to stop the reproduction of cancer cells by stopping abnormal protein synthesis. High-throughput screening for this drug was done in 1994 and in 2001 Bayer pharmaceuticals discovered it initially by a biochemical assay using RAF kinase in which screening of 200,000 different compounds in order to categorize active molecules to counter active RAF kinase. After three testing trials, its anti-angiogenic effects against cancer were discovered starting the process of stopping the synthesis of new blood vessels. Another discovery Maraviroc which is an HIV entrance inhibitor by slowing down the process, thus, preventing the entry of HIV in human cells as well as in the treatment of cancers by blocking the metastasis of cancer cells when the cancer cells start spreading to different parts of the body from its port of origin. High-throughput screening of Maraviroc was done in 1997 and in 2005; it was finalized by Pfizer's global R&D team.

Omics research under which high-throughput biology or screening plays as one side—the link between large scale biology (under which genomic, transcriptomic, proteomic comes under), technology, and researchers. HTS in cell biology has a crystal clear focus on cells and different methods that are required in order to access the cell including imaging, microarrays for studying gene expression and genome-wide screening. These methods are automated with large scale potential without altering the quality in terms of the result as well as the sample.

Due to the presence of big data in life sciences, it is very important to run experiments in automation making large-scale recurrence practical. For an instance, 21,000 genes out of which some may be responsible for cell functioning as well as some for disease in the human genome. So in order to make an idea about how a single gene works and how it interacts with the rest of the genes, in which genes are involved in and located at, thus, such methods capturing the whole procedure from cell to genome. With the decrease in the cost of High-throughput Sequencing (HTS) experimentation, it has been brought within the capacity of small laboratories, providing facilities like production of high-dimensional data sets with HTS generating 100 Gb of data in 24 h. With big data, processes such as designing, pre-processing followed by a downstream study of HTS data is noteworthy (Fig. 7.2).

Furthermore, a number of challenges are present includes collection of sample and control of quality, choosing HTS method as per requirements followed by this assimilation of various data sets from different platforms and other technologies. HTS data itself produces challenges in silico and computationally called "Data Deluge", now the emphasis is more on storing the data, accessing that data, and further analyzing that data successfully. Apart from this, there are data control and patient privacy association as well, due to the pace of alteration by the application of HTS for clinical purposes (Fig. 7.3).

7.3 Genomics

7.3.1 What is DNA?

Deoxyribonucleic acid (DNA) contains all the guidelines required to expand and guide the functioning in every organism. DNA has two strands in twisting fashion

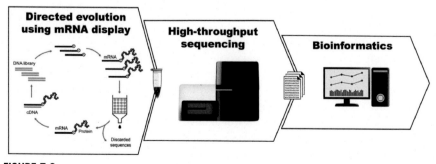

FIGURE 7.2

General representation of isolation of sequences of interest from in vitro experimentation, HTS and further study of the sequencing data.

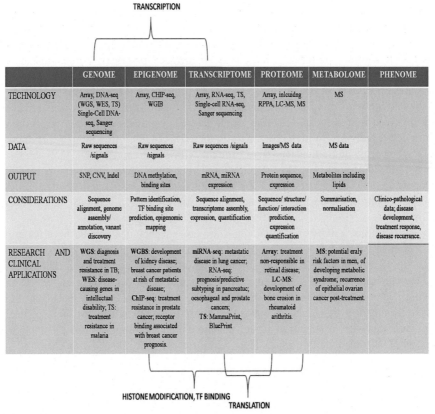

	GENOME	EPIGENOME	TRANSCRIPTOME	PROTEOME	METABOLOME	PHENOME
TECHNOLOGY	Array, DNA-seq (WGS, WES, TS) Single-Cell DNA-seq, Sanger sequencing	Array, CHIP-seq, WGIB	Array, RNA-seq, TS, Single-cell RNA-seq, Sanger sequencing	Array, inlcuidng RPPA, LC-MS, MS	MS	
DATA	Raw sequences /signals	Raw sequences /signals	Raw sequences /signals	Images/MS data	MS data	
OUTPUT	SNP, CNV, Indel	DNA methylation, binding sites	mRNA, miRNA expression	Protein sequence, expression	Metabolites including lipids	
CONSIDERATIONS	Sequence alignment, genome assembly/ annotation, vanant discovery	Pattern identification, TF binding site prediction, epigenomic mapping	Sequence alignment, transcriptome assembly, expression, quantification	Sequence/ structure/ function/ interaction prediction, expression quantification	Summarisation, normalisation	Clinico-pathological data; disease development, treatment response, disease recurrance.
RESEARCH AND CLINICAL APPLICATIONS	WGS: diagnosis and treatment resistance in TB; WES: disease-causing genes in intellectual disability; TS: treatment resistance in malaria	WGBS: development of kidney disease; breast cancer patients at rish of metastatic disease; ChIP-seq: treatment resistance in prostate cancer; receptor binding associated with breast cancer prognosis.	miRNA-seq: metastatic disease in lung cancer; RNA-seq: prognosis/predictive subtyping in pancreatic, oesophageal and prostate cancers; TS: MammaPrint, BluePrint	Array: treatment non-responsible in retinal disease; LC-MS: development of bone erosion in rheumatoid arithritis.	MS: potential eraly risk factors in men, of developing metabolic syndrome; recurrence of epithelial ovarian cancer post-treatment.	

FIGURE 7.3

Summary representation of omics theory related to different techniques, information, result, methodical concern as well as investigation and clinical purpose. Genome to transcriptome to proteome to metabolome is represented.

forming a double helix. Each strand has four bases: adenine (A), thymine (T), guanine (G), and cytosine (C), in which A always makes a double bond with T furthermore G always makes a triple bond with C. The arrangement of As, Ts, Gs, and Cs helps in guiding the information packed in DNA for bodily functions and various mechanisms.

Genomics: Genetics involves the study of hereditary or the transmission of different characteristics of an organism from one generation to another. It involves studying specific and restricted genes which have certain known important function such as guiding body's development, diseases and drug response.

Genomics involves the study of an organism's whole genome. With the help of a higher level of computing and algorithms called bioinformatics which is used for analyzing the genome, thus, researchers study significant quantities of data to come across variations that affect our health, diseases, and drug response. For example, humans going through all 3 billion base pairs in DNA for ∼23,000 active genes. Advancement in DNA sequencing and computational biology in the last few decades has led to this new field of genomics with respect to a much older field of studying genetics (Fig. 7.4).

DNA sequencing: determination of the actual order of bases in a DNA strand. There are different types of sequencing methods such as in sequencing by synthesis, DNA polymerase enzyme generates a new DNA strand from parent strand. The enzyme adds fluorescently tagged nucleotides into the new DNA strand, an indication is obtained from each nucleotide that has been previously tagged, and thus, the signal is detected. Each of four nucleotides produces a different signal reading up to 125 nucleotides successively and billions at a moment Thus, DNA sequencing is used for various research works including genetic variations or mutations involved in the development of the disease. Variations are as small as deletions, substitution or insertion, or deletion of a large number of bases in a row (Mitchelson, 2005).

Human Genome Project: Completed in April 2003 (Powledge, 2003), the NIH led the HGP which gave sequencing results of human genome which was freely accessible public databases. Using for studying the genetic variations which in turn lead to the high risk of specific diseases like cancer. More work in order to understand the functioning of the genome and genetic source for numerous health and disease.

The commonly used technologies and tools for functional genome analysis: Studying variants present within both coding and non-coding in genome in a wide range from a change in single nucleotide to large aberrations in the chromosome which can be easily visible using a microscope, causing an enormous impact on the functioning of the gene. They can be beneficial for example, SNP with no such negative impact on the phenotype or on the other hand, pathogenic for example,

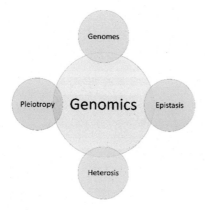

FIGURE 7.4

Genomics studies the relationship between the genomes of all organisms and intragenomics.

a nonsense variant that results in the different disorders and diseases largely dependent on the type of variant and locus. Different genetic methods and tools for the detection of variants are there. In this segment following topics are discussed:

1. DNA microarray
2. DNA-sequencing:
   2.1. Whole Genome sequencing (WGS)
   2.2. Whole Exome sequencing (WES)
   2.3. Targeted Genomic Sequencing (TS)
3. Single-Cell DNA-sequencing

7.3.2 DNA microarray

Researchers are familiar with that a mutation in a gene may lead to an onset of a certain disease. Mutations can occur anywhere because most large genes have many regions, thus, making it difficult to identify mutations (Fig. 7.5). For example, 60% of hereditary in breast as well as ovarian cancers occurs owing to mutations within the BRAC1 along with BRAC2 genes alone in BRAC1 has 800 different types of mutations. The DNA microarray is a tool for determining DNA of an individual is mutated in genes such as BRAC1 and BRAC2. The microarray chip consists of short, synthesized thousands of ss-DNA which as one makes the gene of the

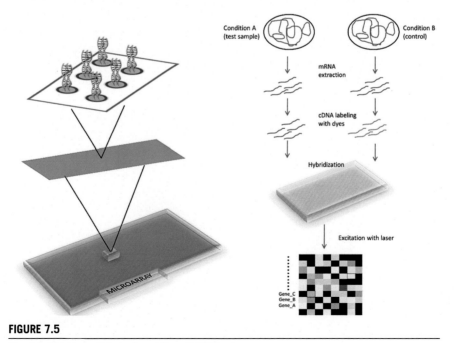

FIGURE 7.5

Graphical representation of workflow of DNA microarray.

query, and to gene variants that are part of the human population (Gresham et al., 2008).

Mechanism of DNA microarray: In order to check mutation of a particular disease, first, sample containing DNA is isolated from the blood of an individual and from the control sample which is normal for that particular gene. Followed by this, denaturation of DNA samples occurs, thus, separating the complementary double strands and long fragments are cut into smaller fragments, each fragment tagged fluorescently using green dye, and control with red dye. On the chip, both tagged DNA are placed on the chip for hybridizing to the synthetic DNA. If sample does not contain the mutation in the gene, equally red tagged and green tagged samples attach with sequence without mutation. If the patient contains mutation, patient's DNA does not attach properly to synthetic DNA sequences although attach to the DNA sequence having mutation on the chip.

7.3.2.1 Application of DNA microarray

- Gene expression analysis—From the cell of interest, RNA extraction is done and it's labeled either directly with labeled complementary DNA or T7 RNA promoter tagged with cDNA which is changed to cRNA. Different methods of cDNA or cRNA labeling such as, during synthesis nucleotides which are fluorescently labeled are incorporated, biotin-labeled nucleotide which are afterward stained with streptavidin which is fluorescently labeled, modified nucleotide which is later tagged with a fluorescent label, and a range of signal amplification methods. The most commonly used methods for labeling cRNA or cDNA is fluorescently labeled nucleotides or nucleotide labeled with biotin.
- Transcription factor binding analysis—Chromatin immunoprecipitation in arrangement with microarrays has been used for determining the transcription factors binding sites. TFs are cross-linked with DNA using formaldehyde, thus, DNA fragmentation occurs. The TF of interest remains bound with DNA, which is isolated using antibody against particular TF or through labeling TF using peptides used for affinity chromatography, for instance, an HA-, FLAG-, HIS-tag.
- Subsequent to refinement, DNA is removed as of TF which is further amplified using PCR, labeled, as well as hybridized with the array. This technique is usually called "ChIP-chip" for on the microarray.
- Genotyping—For SNP genotyping, microarrays have been in use widely. There are other alternative approaches for detection of SNP but allele discrimination through hybridization; allele-specific expansion along with ligation to a barcode oligonucleotide are most commonly used or other approaches which include particular nucleotide expansion reaction the arrayed DNA extended along with SNP.

Limitations of microarray

High throughput technologies based on hybridization are comparatively inexpensive but do have a number of limitations including:

i. Dependency on already existing knowledge about the sequence of the genome
ii. High background noise level due to presence of cross-hybridization
iii. Incomplete variety of recognition because of both background as well as saturation signs
iv. Comparison of expression intensity between diverse range of experiments can be a tedious task, requiring complex normalization mcans
v. Data irregularity particularly for genes having low expression levels
vi. Provides zero knowledge about protein's level of expression and functioning

DNA microarray analysis using bioinformatics tools

a. Qspline used for Affymetrix array data normalization as well as spotted array data is accessible in the Bioconductor: affy package.
b. ClustArray used for array data clustering.
c. OligoWiz used for spotted arrays for designing oligonucleotide probes.
d. ProbeWiz used for spotted arrays for designing PCR primers.
e. Promoter used in the prediction of promoter sites in a vertebrate.

Other tools such as: ArrayExpress, ArrayTrack, BASE, dchip, EzArray, GeneX etc.

R packages for analysis: Affy, PLIER, LIMMA, sihPathway, org.Mm.eg.db.

7.3.3 DNA sequencing

1.1. WHOLE GENOME SEQUENCING (WGS): a complete process used for complete genome analysis. The knowledge which is present at the genomic level is very important for determining inherited disorders, specifying mutations responsible for the progression of cancer, and tracking disease outbreaks such as the Ebola virus, etc. With rapid decrease in the cost of sequencing and the capacity of producing big data with present sequencers making WGS a great tool for research in genomics. With scalability, the flexibility of NGS machinery commonly for human genome sequencing and further making it useful for sequencing a wide range of species such as plants, livestock, and disease-causing microorganisms (Logsdon et al., 2020). WGS was originally executed for the human genome by using Sanger sequencing which took more than a decade and $1 billion. Now with a newer technology called "NGS" with HTT and ability of huge parallel sequencing using both DNA as well as RNA which is very rapid and cheaper than Sanger sequencing with the cost of $1000. For laboratory identification in metagenomics (studying and sequencing microbial genomes), for public health inspection during outbreaks, for example, in cases of *E. coli, Campylobacter j., Legionella p., and Mycobacterium t.* disease at global and local epidemics due to influenza, Zika, Ebola, and now Corona viruses. Used in tracking down the source and spread in order to help in controlling and preventing it at any cost (Fig. 7.6).

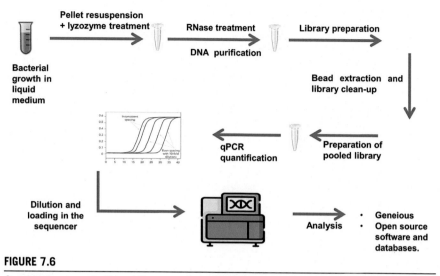

FIGURE 7.6

Graphical summarization of the procedure of WGS.

Technologies used for sequencing failed to sequence the entire human genome at once. As a result, the genome is broken down into smaller DNA fragments, followed by sequencing, and then using bioinformatics tools these sequences are put together accordingly.

7.3.3.1 Clone-by-clone process

This technique in which the genome is fragmented into smaller parts, copied and introduced into bacteria, which grows in order to produce identical copies called "clones" enclosing around 150,000 bps of the genome which is a sequence of interest. Further, each inserted DNA in the clone is again fragmented into smaller fragments with overlapping chunks of 500 bps, followed by sequencing and used for reassembling the clone. This method was used for human genome sequencing using the Sanger method although it took the time of a decade and was extremely costly but was a reliable process.

7.3.3.2 Whole-genome shotgun process

Shotgun sequencing is a technique that breaks DNA into smaller random fragments for sequencing and reassembly. These are used for cloning into the bacteria for increasing the number and isolated for further sequencing. Because of random fragmentation, these overlapping sequences aids in reassembling into the original DNA sequence of order, this was originally used by Sanger sequencing and now used in NGS for rapid genome sequencing at low cost. Although, this method is more efficient for smaller reads and genomes with less repetitive regions for reassembling based on regions of overlapping and requires reference genome as well as computational approach for reassembling.

7.3.3.3 Assembly of sequencing reads

The varying length of DNA fragments are sequenced, the resultant is put back together, which is known as "assembly". Two common methods are: de novo assembly—this method is done first by identification of overlapping areas in DNA sequences, aligning them, and placing them back together to form genome without comparing it with any kind of reference genome. This is used for sequencing unknown, new organisms. Moreover, this method gives fewer biased results in comparison to a reference genome.

Whereas, in mapping to a reference genome, in this method, for assembly, the genome uses another reference genome for aligning new sequencing data. This method is quite easier and needs less contagious reads, but difficult for new sequences. The result depends upon the type of reference genome used and provides better detection of SNPs, thus, multiple reference genomes have been produced for different races/ethnicities by multiple institutes and companies for studying SNPs known to a particular race and ethnicity.

Application of WGS:

- Mutation frequencies—WGS has been successfully used for studying frequency of mutation in human genome. The rate of mutation is 70 novel mutations for every generation of humans that's a parent to child in the whole genome. In coding regions of the genome, 0.35 mutations are there which would alter the protein sequence among generations. Because of genome instability, the frequency of mutation is much higher in cancer depending on a number of factors such as age, UV exposure, habits like smoking, tobacco.
- Genome-wide association studies—In research, WGS is widely used for GWAS—a project planned toward establishment of a relationship between genetic variant or variants with a disease or several new phenotypes.
- Diagnostic use—for various infectious outbreaks or neurological diseases such as Alzheimer's disease.

7.3.4 Whole exome sequencing (WES)

An NGS technique in which coding regions for proteins of the genome are sequenced. Exome part of the human genome is less than 2% consisting of approximately 85% disease-related variants making it cost-efficient when compared to WGS (Koch, 2021). WES uses exome enrichment which powerfully identifies coding variations across a wide array of applications such as genetic disease, cancer analysis, and population genetics (Fig. 7.7).

Applications of WES: WES is a useful technique for clinical application because it uses coding regions of the genome for determining variants in exon regions responsible for the disease. There has been a huge increase in WES data; it has been successfully used in determining gene association with Mendelian phenotype, miller syndrome, and various complex disorders. Ever since 2011, it is used as a diagnostic tool for clinical genetic laboratories. It has been part of 1000 genome

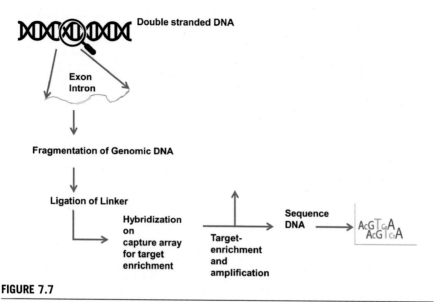

FIGURE 7.7

Graphical summarization of the procedure of WES.

projects, great efforts by NHLBI—GOESP, and ExAC in order to make a catalog of rare variants present in the population for identifying diseases. Such efforts bringing patients with more personalized medicines and treatment approaches as per their own exome.

TARGETED GENOME SEQUENCING (TS): For analyzing specific mutations of an individual's sample then targeted gene sequencing is a useful tool (Burgess, 2020). Panels consist of a selected array of genes or regions of genes having a known or questionable relationship with a disease or particular phenotype. TS having huge scalability pace as well as capacity for estimating targeted genes with a large multiplicity of samples in parallel which in turn saves time and cost in comparison to multiple separate tests. TS is based upon deep sequencing and generate a more manageable small data set and requires small amount of input in comparison to WGS, thus, making its analysis easier and faster (Fig. 7.8).

Bioinformatics Tools used for data analysis of WGS, WES, and TS:

- Read Alignment—BWA, Bowtie
- Annotation—Annovar (Qiagen), Variant effect predictor (Ensembl), SNPsift and SNPeffect, Sift4G,
- NCBI Variant Annotation
- Visualization—NCBI Variant Viewer, UCSC Genome Browser, ExAC Browser, Personal Genome Browser, 3D Genome Browser
- Data-warehousing—ClinVar, dbSNP, GWAS Catalog, GWAS Central, Cancer Atlas, RefSeq, PANTHER

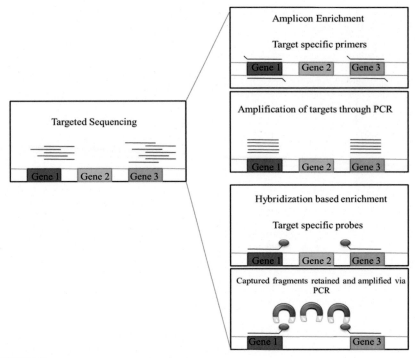

FIGURE 7.8

Graphical summarization of the procedure of TS.

- Analytics—Genome Analysis Toolkit, MuTect, ASEQ, Halvade-RNA, GT-WGS, KaryoScan
- AI based Analytics—Exomiser, DeepVariant, Deep Genomics, Lifemap Science

7.3.5 Single cell DNA-SEQ (sc-DNA-seq)

Another NGS technique called Single-cell DNA sequencing (scDNA-seq) is a resourceful and scalable method for studying genetic heterogeneity in organisms multi cellular (Gawad et al., 2016). WGS for bulk tissues is used for identifying somatic mutations, although, it has limited sensitivity for mutations having low proportion in cells (Fig. 7.9).

Process of sc-DNA-seq:

1. Isolation of single cell
2. Whole genome amplification of isolated cell
3. Library preparation
4. Sequencing
5. Data analysis

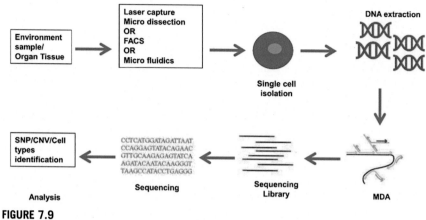

FIGURE 7.9

Graphic representation of workflow of single cell genome sequencing.

Application of single-cell DNA seq:

- For Tumor cells: The sc-DNA-seq method overcomes all the limitations faced normally in sequencing techniques. The mapping of tumor cells and the micro-environment in which tumor cells grow is done by means of detection of heterogeneity of tumor cells, it clarifies the cell cluster and finds specific markers, which eventually explains tumorigenesis and metastasis. Thus, sc-DNA-seq has been used for studying tumor cells and finding different diagnostic and treatment methods.

- For nervous system: Due to the unique variations within nerve cells which cause differences among each neuron in the nervous system. Due to this heterogeneity involved in these neurons, it's complicated to learn the brain circuits and resolving these issues related to connectivity. But, the sc-DNA-seq technique studying different stages of nerve cells and drawing each detail on a map of single-cell in order to determine diverse kinds of neurons and their connection with brain.

- For reproductive and embryonic medicine: The sc-cell DNA-seq technique on single-cell stage sequences and quantifies the genome in germ cells as well as embryonic cells, eventually assist in the understanding processes such as screening, diagnosis, and treatment of diseases related to reproduction and genetics.

- For immunology: due to great heterogeneity among immune cells again analysis with sc-DNA-seq aids in studying these cells for better diagnosis and treatment for various diseases.

- For digestive system and urinary system: Studying digestive and urinary system cells using sc-DNA-seq aids in mapping the machinery of cells such as intestinal cells sustaining homeostasis plus countering to microorganisms which are pathogenic is clarified.

7.4 **Epigenomics**

Epigenetic means "on genetic" sequence, any process by which the gene activity is altered without making any change in the DNA sequence and leading to alternations which are passed to daughter cells (Allis and Jenuwein, 2016). Epigenetic means studying a single locus or pair of loci, whereas, epigenomics means the whole study of epigenetic modifications in the entire genome. Epigenomics includes the investigation of phenotypic or quality articulation changes brought about by inheritable systems autonomous of DNA succession, e.g., DNA methylation and histone post-translational changes. Likewise, noncoding RNA adjustments can be delegated epigenetic components, as they have been appeared to intervene epigenetic DNA and histone modifications. Epigenetic marks are progressively being perceived as significant components hidden phenotypic variety, natural cycles, irritation, and sicknesses for example, disease, and enormous scope epigenome projects have been set up trying to make reference human epigenome maps. High-throughput procedures empowering epigenetic investigation on a genome-wide scale incorporate, e.g., shotgun bisulfite sequencing and pyrosequencing to dissect DNA methylation, also, genome-scale chromatin immunoprecipitation with antibodies perceiving explicit histone changes or DNA methylation, trailed by microarray examination or high-throughput sequencing. 146, 147 Alternative strategies to inspect DNA methylation is differential methylation hybridization, which envelops methylation-touchy DNA limitation followed by microarray examination of the limitation ensured, hyper-methylated DNA, 148 dab cluster methylation investigation of bisulfite-treated DNA, 149 and base-explicit cleavage joined with grid helped laser desorption ionization/season of flight. However, epigenetics is cell-ward and exposed to ecological components, asking infection explicit investigations to uncover epigenetic changes related to the illness. Contrasted and, for instance, malignant growth research, the cardiovascular field is as yet in its early stages of epigenetic research. Starting considers have tended to vascular aggravation related epigenetic changes on a worldwide level (inspecting worldwide changes in epigenetics without linkage to explicit quality advertisers) or have zeroed in on explicit qualities. For instance, genomic DNA in human atherosclerotic injuries has been demonstrated to be hypo-methylated, while conflicting outcomes have been acquired infringe blood lymphocytes from patients with cardiovascular sickness.

The classical explanation of Transgenerational epigenetic inheritance includes environmental factors which strike pregnant person (Fo) cause direct effect not only in F1 generation, but also on germ cells which signifies the F2 generation. Thus, only modifications during F3 can be seen entirely because of epigenetic inheritance. In male germ line can have an effect only in one generation, with detectable epigenetic inheritance already in F2 (Fig. 7.10).

Different methods for studying Epigenomics are:

1. ChIP-seq
2. Whole-Genome Shotgun Bisulfite Sequencing

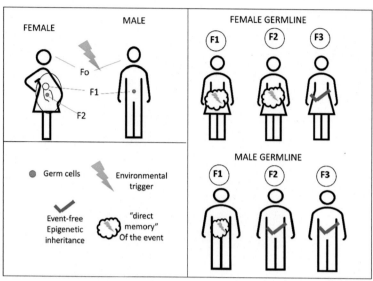

FIGURE 7.10

Graphical representation of trans-generational epigenetic inheritance.

7.4.1 ChIP-seq

First ChIP was carried out along with DNA microarray (ChIP-chip) of DNA-binding proteins in genome-wide mapping. ChIP then sequencing is called ChIP-seq which is a great method used for analyzing histone modifications, proteins, variants, chaperons, nucleosomes, chromatin regulators, TFs, and cofactors throughout the genome (Nakato and Sakata, 2021). Usually, ChIP-seq methodology requires cross-linking by formaldehyde in order to stick DNA-protein interactions, chromatin is fragmented into 100–300 bps using sonication or enzymatic digestion, followed by this, immuno-enrichment of the target of interest using specific antibodies against it, then cross linking is overturned also ChIP DNA is obtained. For PCR amplification, first adaptor ligation afterward single or paired terminal sequencing is carried out. ChIP-seq is one of the main techniques used for producing reference epigenome maps in a number of large epigenomics projects. ChIp-seq has been used for disease epigenetic studies in terms of epigenetic alternations in cancer and noncancerous diseases and for precision medicine development (Fig. 7.11).

Application of ChIP-seq:

ChIP-seq is used for determining the association between transcription factors and chromatin-associated proteins that manipulate the phenotype mechanism of an organism. How proteins associates with DNA in order to control gene expression for completely understanding biological processes and diseases. This epigenetic data is corresponding to genotype and analysis of expression. ChIP-seq technology is presently seen mainly as a substitute for ChIP-chip which involves a hybridization array.

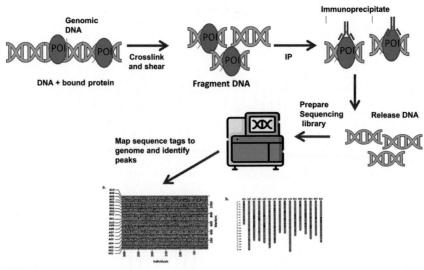

FIGURE 7.11

Graphical representation of ChIP-seq.

Bioinformatics tools used for analyzing ChIP-seq data:

- Short-read aligners—BWA, Bowtie, GSNAP
- Peak Callers—MACS, PeakSeq, ZINBA

7.4.2 **Whole-genome shotgun bisulfite sequencing (WGSBS)**

In shotgun sequencing, the whole genome is fragmented short which are sequenced in a parallel fashion. Then these fragments are aligned using computer software in order to reconstruct the complete sequence (Miura et al., 2012). Most NGS depends on this, whereas, NG-bisulfate-seq method in which genomic DNA is treated using sodium bisulfite, then ligation of adapters to both the ends of the fragments for amplification and followed by sequencing. In WGSBS, bisulfite decreases the genome from 4 to 3 bases making primer designing for whole-genome difficult. The library is selected according to size and separated using NG-stage. WGSBS provides highest resolution and 5 mC level of analysis. Although, WGSBS is quite expensive and demands a large amount of DNA samples for each run due to because of bisulfite fragmentation of DNA.

Extraction of genomic DNA from tissue afterward fragmented and both ends are arranged for universal adapter ligation. Main feature of this process is sodium bisulfite alters un-methylated C into U, whereas, after sodium bisulfite treatment methylated cytosines are left unchanged and thus, in a population of cells, levels of methylation can be analyzed at a single nucleotide by counting the cytosines number for ratio of entire amount of examine at that particular nucleotide. Following

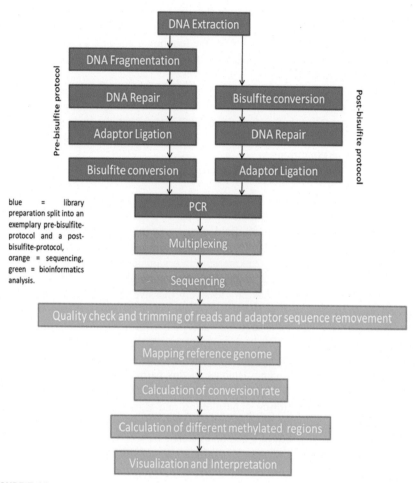

FIGURE 7.12

Flowchart of the core steps in WGSBS.

acquiring 100 bp reads using Illumina as well as bioinformatics investigation permits determining methylation on single-nucleotide resolution throughout the DNA sequence (Fig. 7.12).

7.5 **Transcriptomics**

Transcriptomics is the study of genome-wide RNA expression, using microarray technology. It is one of the most used techniques for analyzing disease mechanisms and it has widely used in studying expression profiles in human, rat, mouse, pig, and monkey of atherosclerotic-prone vessels (Quake, 2021). The transcriptome is total transcripts at a particular stage of development. Analyzing the Transcriptome is important in order to understand basics of the functional genome as well as enlightening the constituents in cells, tissues at the molecular level for more understanding of disease development.

The vital points of transcriptomics are: listing whole transcript species, including mRNAs, non-coding RNAs, and sRNAs; to make the transcriptional mechanism of genes, as far as their start sites at 5′ and ends at 3′, type of splicing, and rest of PTMs; and evaluating varying expression intensity of every transcript throughout developmental and diverse conditions.

Different methods for studying Transcriptomics are:

7.5.1 RNA-seq

As a significant utilization of NGS, RNA-seq is growing quickly from last few decades and has become a significant methodology for transcriptome examination and quantitative investigation of gene expression in organisms (Wang et al., 2009). The advancement of high-throughput sequencing innovation set apart by NGS shows the following qualities: the NGS has progressively massive identification throughput, small detection time as well as low detection cost. The third-gen sequencing stage has worked on long fragment sequence analysis, with a broad spectrum of flux and detection.

Utilizing the RNA-seq to investigate sequencing of the transcriptome of the organism can supplement the gene database of this particular species, get countless expressed sequence tags (ESTs) data, and find some new functional genes, which is advantageous to the resulting gene cloning and significant molecular markers improvement. RNA-seq can likewise examine the worldly and spatial expression of a particular tissue or genes of a cell and investigate some obscure sRNAs, which has been broadly utilized in disease detection, drug screening and drug system and so on RNA-seq innovation has numerous advantages, for example,

1. High resolution: RNA-seq can precisely recognize single bases, for example, issues such as background noise and cross-linking due to fluorescence signal can be easily avoided.
2. High throughput: Through the transcriptome sequencing technique, countless base arrangements can be acquired, which can essentially cover the entire transcriptome.
3. High sensitivity: The rare transcripts which are really low are recognized by the RNA-seq strategy.
4. Convenient to use: This innovation can be utilized to examine the entire transcriptome of different species and needn't bother with the reference genome or designing specific probes prior to sequencing. Thus, RNA-seq can easily analyze the entire transcriptome.

Initially, long-RNAs are transformed into a library of cDNA fragments. Sequencing adaptors indicated in color blue are for labeling every cDNA fragment and a small sequence is acquired against every cDNA utilizing an HTS technique. The subsequent sequence are lined up against transcriptome, and differentiated into: exonic, junction, and poly-A end-reads. These are utilized for creating a base-resolution profile designed for all genes, yeast ORF enclosing one intron (Fig. 7.13).

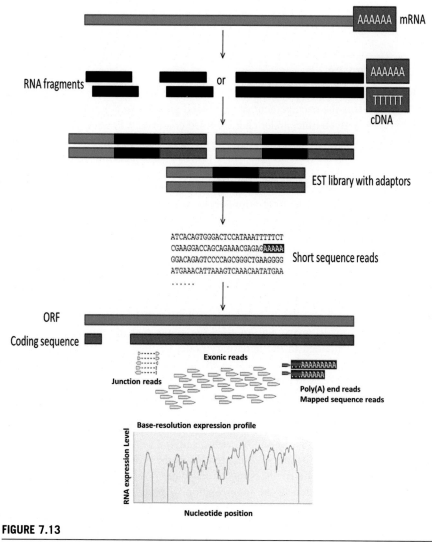

FIGURE 7.13

Flowchart of the core steps in RNA-seq.

RNA-seq Visualization and Advanced Analysis Tools - Software and Resources:

Visualization: Web-based: expVIP, spongeScan, ASCOT, DEIVA, Jbrowse, ATGC transcriptomics, RNASeqExpressionBrowser, ReadXplorer, SaVanT and many more.

Stand-alone: Cascade, omicplotR, TRAPR, SpliceDetector and many more.

Gene fusion: Subread, SOAPfusion, STARChip, FusionHunter, and many more.

Pipelines: RSEQREP, DRAP, TRAPR, NGScloud, RUM, FRAMA and many more.

Other: MINTmap, kissDE, SimBA, KAPAC, GDCRNATools, coseq, bcSeq (Duke), CPSS, and many more.

7.6 Proteomics

The whole arrangement of proteins that is or can be coded by an organism' cell at a specific point. It is the arrangement of expressed proteins in a given sort of cell or living being, at a given time, under characterized conditions. Proteomics is the analysis of the proteome. The term proteome has been applied in different biological systems. A cell proteome are the group of proteins in a specific cell type under particular environmental surroundings, for example, during hormone stimulation. It can likewise be valuable to think about an organisms' whole proteome, which can be conceptualized as the total arrangement of proteins from the entirety of the different cell proteomes. This is generally what could be compared to the genome. The expression "proteome" has likewise been utilized to allude to the group of proteins in certain sub-cellular biological frameworks. For instance, the entirety of the proteins in an infection can be known as a viral proteome. The entirety of the proteins in a mitochondrion makes up the mitochondrial proteome which has created its own field of study metaproteomics.

The long-standing worldview in science is that DNA incorporates RNA, which integrates protein. Customary way of thinking states that the outline for how to collect a cell is contained in the hereditary code, yet understands that the blocks and mortar utilized in the structure cycle are transcendently proteins. In this way, proteins are the particles in cells that are straightforwardly answerable for the upkeep of right cell work, and thus the reasonability of the creature that contains the cells. Lately, the concurrent investigation of the entire scope of proteins communicated in a cell at some random time has become a zone of extraordinary premium. This has prompted the order of another sub discipline of protein science known as 'proteomics', where a proteome is characterized as the protein supplement communicated by the genome of a living being or cell type.

The instruments utilized in the investigation of proteins, nonetheless, actually linger behind the comparable to devices utilized in the examination of DNA and RNA. It is moderately effortless to embrace the recognizable proof and evaluation of various DNA or RNA particles in a solitary investigation utilizing an exhibit arranged from a solitary beginning example. This should be possible utilizing such strategies as DNA chips and cDNA microarrays, differential presentation PCR, and sequential investigation of quality expression. It is essentially impractical to play out similar sort of investigations at the protein level utilizing two-dimensional gel elec-trophoresis (2DE), which is the current generally acknowledged innovation here notwithstanding the way that it experiences a few significant deficiencies. In talking about scientific techniques to be utilized in proteome examination, thought should be given to the way that the quantity of proteins communicated at one time in a given cell framework is commonly in the large numbers or several thousands. Any endeavor to order and recognize these proteins at the same time should utilize tech-niques that are pretty much as fast as conceivable to empower fulfillment of the ven-ture inside a sensible time span. Along these lines, a glorified proteomics innovation would comprise of a blend of the accompanying highlights: high affectability, high output, capacity to separate various changed proteins, as well as the capacity to examine all the proteins present in a sample (Meissner et al., 2022).

DIFFERENT METHODS FOR STUDYING PROTEOMICS ARE:

1. Reverse phase protein microarrays (RPPA)
2. Mass Spectroscopy (LC-MS/MS)

7.6.1 Reverse phase protein microarrays (RPPA)

A protein microarray that permits estimation of protein expression intensity in countless biological conditions at the same time in a quantitative way when anti-bodies are free. Actually, infinitesimal measures of (a) cell lysates, from whole cells (b) fluids extracted from body, for example, a serum, urine, salivation, and so on, are placed on individual points on microarray further incubated with particular antibody to distinguish the expression levels of the protein of interest checking numerous ex-amples. On an individual microarray, as per the aim, can compel thousands of tests which are imprinted in duplicates. Identification is executed utilizing either primary or secondary tagged antibodies using chemiluminescent, fluorescent, or colorimetric techniques. The obtained array is evaluated.

Multiplexing is accomplished by multiple probing of arrays that are spotted with the same lysate with various antibodies at the same time and can be used as a quan-titative measure.

APPLICATION OF RPPA: RPPA progressively is utilized for detecting deregulated signaling pathways of various disease tissues. RPPA is a strategy used for deciding whether multi-omics therapy at molecular level enhances the clinical route of patients having metastatic breast disease estimated using the growth mod-ulation index, it is determined like proportion of period of instances in treatment and development of primary tumor/metastases. Other than profiling flagging pathways or whole organizations in disease tissues related to human, normally used techniques for approval of MS found biomarkers is RPPA.

It is utilized for deciding possible variances of protein intensity. The information permitted order in protein and phosphor-protein intensity throughout the preanalytical stage in: (1) predictable stable; (2) predictable stable; (3) unpredictable. As the majority of phosphor-proteins had a place with the last gathering, the creators suggest tissue obsession or adjustment after specimen assortment immediately. Consequently, tissue obtainment rules ought to be adjusted. RPPA has likewise been utilized to investigate the complexity of proteins within a primary tumor as well as the lymph node metastases of a similar patient. These investigations uncovered a huge complexity of a subset of proteins inside a tumor or between primary tumor and metastases, proposing molecular determination utilizes numerous tumor tests from diverse areas instead of examination of one single test ought to be imagined.

BIOINFORMATICS TOOLS FOR RPPA DATA ANALYSIS: RPPAware, Supercurve, Normacurve, Rppanalyzer, Rppapipe, Reverse Phase Protein Microarray Analysis Suite, RPPAML/RIMS, and many more.

7.7 Metabolomics

The wide-scale analysis of small molecules or metabolites is called Metabolomics; these are found in the cell, tissue, biofluids of an organism (Johnson et al., 2016). Together these metabolites and their interactions with each other in a living system are described as metabolome. Genomics in which the DNA is studied and transcriptomics in which RNA is studied as well as its expression; Metabolomics study of metabolite products, all influenced because of genetic and environmental conditions. Metabolomics is a very important topic of research due to different metabolites and their interaction with each other, not like the rest of the omics; it directly explains the fundamentals of biochemical activities inside the cells/tissues. Hence, Metabolomics explains molecular phenotype best.

Practically speaking, metabolomics presents a critical logical test in light of the fact that, not at all like genomic and proteomic techniques, it intends to quantify atoms that have dissimilar actual properties (e.g., going in extremity from water-dissolvable natural acids to nonpolar lipids. Likewise, far-reaching metabolomic innovation stages normally take the system of isolating the metabolome into subsets of metabolites—frequently dependent on a compound extremity, basic practical gatherings, or primary comparability—and devise explicit example arrangement and scientific techniques advanced for each. The metabolome is thusly estimated as an interwoven of results from various insightful strategies. As an arising field that has been empowered, at any rate to some degree, by the consistent advancement of scientific instrumentation using new capacities every time, techniques utilized for metabolomics proceed for more advancement. Notwithstanding, a result of metabolomics labs utilizing different strategies that are possibly liable to visit refinement is that singular research centers will in general have interesting techniques and there is similarly barely any standard working methodology normally received across labs. Albeit this variety of advancements is connected to development in the field, it fits potential difficulties when looking at information between research facilities on account of issues like contrasts in the exactness of estimation for selected types of metabolites. What's more, the level of assurance in metabolite ID can fluctuate among techniques, going from metabolite characters thoroughly affirmed utilizing true reference principles to putative recognizable pieces of proof made utilizing reference data sets to signals that stay as "questions." The requirement for normalization in metabolomics has been valued by its experts and has offered to ascend to various activities toward understanding this point, for example, Metabolomics Standards proposal to create rules intended for information announcing; ring tests to evaluate the capacities of a variety of metabolomics strategies as well as labs to get equal outcomes; and easily accessible vaults for metabolomics outcomes and associated metadata, for example, the MetaboLights data in Europe (http://www.ebi.ac.uk/metabolights) and the Metabolomics Workbench (http://www.metabolomicsworkbench.org) in the United States.

7.7.1 Different methods for studying metabolomics

Mass spectroscopy is used for studying both proteomics and metabolomics analysis. In this session, I've discussed both MS technique for proteomics and metabolomics.

Among the tool compartment of methods with which proteins can be explored for an enormous scope, mass spectrometry (MS) has acquired notoriety on account of its capacity to deal with the complexity related to the proteome. Different strategies, for example, 2D gel electrophoresis, two-hybrid analysis, as well as protein microarrays neglect in accomplishing in-depth proteome examination observed using mass spectrometry. The essential utilization of MS in the field of proteomics are classifying protein expression, characterizing various protein relations among each other, as well as recognizing sites of alteration in protein. Utilization of MS in the field of proteomics is not the use of the single application for all analysis but rather a group of methodologies, each having qualities fit for specific requests. In every MS analysis, thought ought to be given to the kind of instrumentation, fragmentation technique, and investigation system most appropriate to an individual example (Fig. 7.14).

MS-based identification of protein and characterization of its functions, first Proteins removed from samples can be examined using bottom-up or else top-down techniques. In the bottom-up methodology, proteins found within complex combinations can be isolated prior to enzymatic fragmentation followed by direct peptide mass estimation obtained or additional to this, the peptide partition along with tandem mass spectrometry. Then again, the proteins can be easily processed into the fragmentation of peptides which are then isolated by chromatography as well as tandem mass spectrometric study. In the top-down methodology, protein complexes are fragmented and isolated into only pure proteins or combination of fewer complexities, trailed with a static mixture of the sample into MS for actual mass estimation of protein. An on-line LC-MS technique can likewise be utilized for huge scope protein cross-examination (Fig. 7.15).

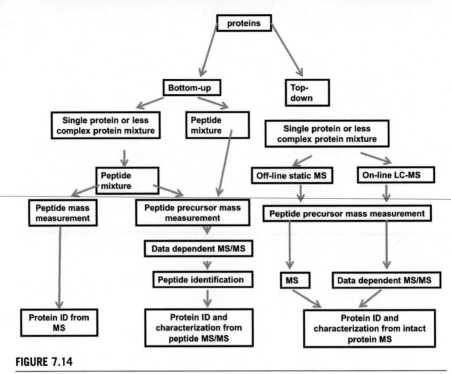

FIGURE 7.14

Flowchart of the core steps in MS for proteomics.

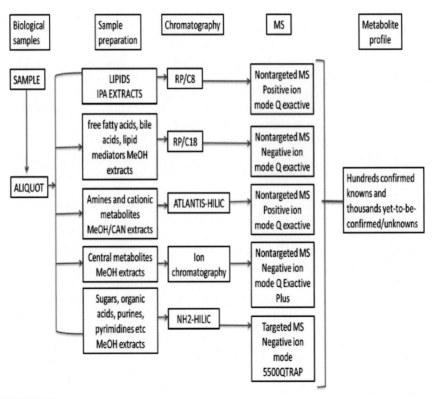

FIGURE 7.15

Flowchart of the core steps in MS for metabolomics.

SOFTWARE TO IDENTIFY AND QUANTIFY PROTEINS USING MS DATA:

- De novo sequencing: PepNovo, PEAKS
- Database searching: SEQUEST, Mascot, X!Tandem.
- Peptide identification: PeptideProphet, Precolator.
- Spectral library: SpectraST, X! Hunter.
- AMT and protein quantification: Viper
- De novo sequencing and database searching: TagRecon, InsPecT.
- Protein identification: ProteinProphet, MaxQuant.
- Protein quantification (ICAT): ASAPRatio
- Differential protein detection using peptide ion current area: MSstats
- Differential protein detection using spectral count: QSPEC

SOFTWARE LIST FOR METABOLOMIC MS DATA ANALYSIS:
metaXCMS, XCMS, XCMS2, MeDDL, MetAlign, MAVEN, centWave, mzMine2, MetabolomeE xpress, Chromaligner, and many more.

References

Allis, C.D., Jenuwein, T., 2016. The molecular hallmarks of epigenetic control. Nat. Rev. Genet. 17 (8), 487–500. https://doi.org/10.1038/nrg.2016.59.

Burgess, D.J., 2020. Complex targeted sequencing in real time. Nat. Rev. Genet. 22 (2), 67. https://doi.org/10.1038/s41576-020-00324-6.

Crick, F., 1970. Central dogma of molecular biology. Nature 227 (5258), 561–563. https://doi.org/10.1038/227561a0.

Gawad, C., Koh, W., Quake, S.R., 2016. Single-cell genome sequencing: current state of the science. Nat. Rev. Genet. 17 (3), 175–188. https://doi.org/10.1038/nrg.2015.16.

Gresham, D., Dunham, M.J., Botstein, D., 2008. Comparing whole genomes using DNA microarrays. Nat. Rev. Genet. 9 (4), 291–302. https://doi.org/10.1038/nrg2335.

Johnson, C.H., Ivanisevic, J., Siuzdak, G., 2016. Metabolomics: beyond biomarkers and towards mechanisms. Nat. Rev. Mol. Cell Biol. 17 (7), 451–459. https://doi.org/10.1038/nrm.2016.25.

Koch, L., 2021. The power of large-scale exome sequencing. Nat. Rev. Genet. 22 (9), 549. https://doi.org/10.1038/s41576-021-00397-x.

Logsdon, G.A., Vollger, M.R., Eichler, E.E., 2020. Long-read human genome sequencing and its applications. Nat. Rev. Genet. 21 (10), 597–614. https://doi.org/10.1038/s41576-020-0236-x.

Meissner, F., Geddes-McAlister, J., Mann, M., Bantscheff, M., 2022. The emerging role of mass spectrometry-based proteomics in drug discovery. Nat. Rev. Drug Discov. 21 (9), 637–654. https://doi.org/10.1038/s41573-022-00409-3.

Mitchelson, K.R., 2005. DNA sequencing. Encyclopedia of Analytical Science, second ed., pp. 286–293. https://doi.org/10.1016/B0-12-369397-7/00683-X

Miura, F., Enomoto, Y., Dairiki, R., Ito, T., 2012. Amplification-free whole-genome bisulfite sequencing by post-bisulfite adaptor tagging. Nucleic Acids Res. 40, e136. https://doi.org/10.1093/NAR/GKS454.

Nakato, R., Sakata, T., 2021. Methods for ChIP-seq analysis: a practical workflow and advanced applications. Methods 187, 44–53. https://doi.org/10.1016/J.YMETH.2020.03.005.

Powledge, T.M., 2003. Human genome project completed. Genome Biol. 4 (1), 1–3. https://doi.org/10.1186/GB-SPOTLIGHT-20030415-01.

Quake, S.R., 2021. The cell as a bag of RNA. Trends Genet. 37, 1064–1068. https://doi.org/10.1016/j.tig.2021.08.003.

Szymański, P., Markowicz, M., Mikiciuk-Olasik, E., 2012. Adaptation of high-throughput screening in drug discovery—toxicological screening tests. Int. J. Mol. Sci. 13, 427. https://doi.org/10.3390/IJMS13010427.

Wang, Z., Gerstein, M., Snyder, M., 2009. RNA-Seq: a revolutionary tool for transcriptomics. Nat. Rev. Genet. 10 (1), 57–63. https://doi.org/10.1038/nrg2484.

Drug informatics

8

8.1 Introduction

The blending of science of information, data, and technology is Informatics. Drug informatics is implementing that data, technology, and drug information in clinical and research settings. Drug informatics is still a young domain in comparison to another medical disciplines. It is a quickly growing discipline that applies medical and health data of computing and information technology. Informatics aims to use technology to assist individuals in performing cognitive activities more efficiently than creating systems to imitate or substitute human expertise. Drug informatics promotes technology as a vital component for a successful organization, analysis, and management of the drug use information in patients. It is linked to data science and characterized as a "medical informatics subset which emphasizes the use of medication data and advancement for enhancing prescription data." Other definitions emphasize the information related to drugs like collection, storage, evaluation, utilization, and distribution of pharmaceutical data (Ou-Yang et al., 2012).

Drug information is extracted from many sources. Primary level sources include a pharmaceutical company or university research lab and clinical observations done at a hospital. Secondary level source for drug data is data created using the data initially obtained for non-clinical objectives, like reimbursement and management of pharmacy benefits. Tertiary-level sources for information on drugs are databases, reference books, journal articles, or clinical pathways based on precisely defined evidence.

The process of drug discovery involves identifying possible novel therapeutic entities using a blend of clinical, experimental, computational, and translational patterns. Although biotechnology has advanced quite a bit and biological systems are now much understandable, drug discovery is still a long, costly, complex, and inefficient process with a high rate of attrition for the discovery of novel therapeutics. It is an innovative process for developing novel drugs based on biological objective information. In the Big Data era, Drug design often depends but not necessarily on bioinformatics and computer modeling techniques. Development and discovery of drugs involve animal models and cell-based preclinical research along with clinical trials on humans, ultimately advancing toward acquiring regulatory authorization to commercialize the medicine. Drug discovery in modern times includes identifying

All About Bioinformatics. https://doi.org/10.1016/B978-0-443-15250-4.00007-1

screening hits, optimizing these hits, and medicinal chemistry to maximize selectivity, metabolic stability, affinity, oral availability, and effectiveness. When a molecule has been identified that satisfies every requirement, drug development will commence before clinical trials. The drug discovery process generally includes setting up a drug design concept, understanding targets (cell tissues, enzymes) character related to disease, and optimizing lead compound using SAR (structure-activity relationship).

Drugs have an essential part in our society, as significant substances in diagnosing, preventing, and treating diseases and improving the "quality of life." They also allow individuals to lead a life with a disease in accordance with their cultural and social background. Whenever a drug is produced or discovered, it significantly affects the quality and overall life.

8.2 Computational drug designing and discovery

The process of drug discovery is exceptionally complicated and requires a multidisciplinary effort to develop effective and financially viable drugs. Computers have an essential part in medical, pharmaceutical, and other scientific research, including discovering a novel drug/compound for improved therapeutic agents (Song et al., 2009). Over the past few decades, there has been an immense increment in computational power and the accessibility to chemoinformatic data, which have enabled computational chemistry and biology methodologies to become an essential instrument for drug discovery. Drug development conventional methods are expensive procedures, extremely time-consuming, with a limited turnover rate of druggable novel chemical entities. The discovery of new therapeutic drugs is made in combination with structural biology and rational medicines. A joint effort by several disciplines is needed to determine new and efficient drugs. So, the CADD center collaborates and works with computational scientists, biophysicists, and structural biologists to develop novel drugs. The CADD methodology helps to streamline the procedure of drug development and discovery through numerous computational methodologies. It employs cutting-edge technology to enhance the time and cost-effective workflow of drug development (Fig. 8.1).

It takes several years, approx. 10–15 years for discovering and developing a drug as it commences with scientific experiments such as specific target receptor determination, disease determination, active compound determination from a mass of compounds, etc. Hence, the scientific community emphasizes minimizing costs and time of drug development without negatively impacting the quality. Many advancements were made during the 1990s by employing high-performance and combinatorial screening methodologies to expedite the development of the drugs. These approaches were extensively accepted as they allowed the quick screening and synthesis of huge libraries possible. However, disappointingly, there was no substantial success, but a somewhat small advancement happened in discovering novel molecular entities. Eventually, the combinatorial approach was introduced, and the term

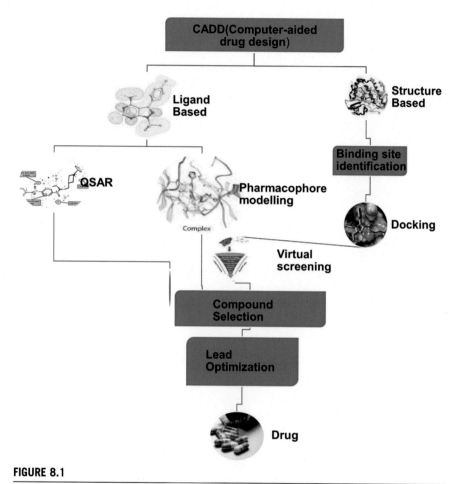

FIGURE 8.1

Flowchart explaining CADD (computer-aided drug designing).

Computer-Aided Drug Design (CADD) came into existence for using the computer in drug development. These advanced tools showed remarkable results, and since then, practically every phase of the drug discovery computational approach has been implemented, like identifying leads, target identification, preclinical data generation, optimization, and development of drugs to the formulation. In drug discovery, CADD has given some of the best results such as - Zanamivir (Neuraminidase inhibitor), Captopril (Angiotensin-converting enzyme inhibitor), Imatinib (Tyrosine-kinase inhibitor), Dorzolamide (Carbonic anhydrase inhibitor), Nelfinavir (HIV-1 protease inhibitors), and Aliskiren (Renin inhibitor). These recent successful researches demonstrate clearly that CADD offers realistic and practical strategies for helping biologists and chemists achieve their objective of developing new beneficial and active compounds while removing toxic, reactive, and inactive compounds.

CADD is often used in the drug development process for three main reasons: firstly, for the substantial compound libraries filtration resulting in smaller subsets with anticipated biological testing activity. Secondly, for the lead compound optimization into optimal pharmacodynamics and pharmacokinetic properties. Last but not least, for designing of new drugs by the approach of de novo designing. These are relatively common approaches and are usually practiced in the development of the drug.

Computational techniques are founded on the concept that any pharmacologically active compound interacts with targets such as nucleic acids and proteins. The main elements governing such molecular interactions among receptor and drug include hydrophobic interactions, molecular surface, hydrogen bond formation, and electrostatic force. These are simply characteristics that are evaluated when the interaction between two molecules is predicted and analyzed. CADD documents every product and streamlines the production process. The substances to be studied might come from several natural sources such as animals, plants, microorganisms, and some synthetic ones. Tested drugs after the test can be refused or approved depending upon the results, such as the presence/absence of carcinogenicity, toxicity, or low efficiency.

CADD can be categorized broadly into two approaches: ligand-based (LBDD) and structure-based drug design (SBDD).

8.3 Structure based drug designing

It uses protein three-dimensional structure data or target (receptor/enzyme) for screening/identifying potential hits, following, synthesis action, biological testing, optimization, and development of novel biologically active compounds. The foremost prime step of SBDD is identifying the target molecule and structure determination. The target identified can be an enzyme-linked to a disease/disorder of interest. The potential molecules are identified based on determining binding affinity, which mitigates the target activity with its inhibition. Therefore, SBDD uses biological target data and discovers potentially novel drugs (Batool et al., 2019). SBDD embodies the significant progress in computational approaches used in statistics, biophysics, biochemistry, medicinal chemistry, and various branches. Scientific and technological advances have led to a considerable number of protein structure predicting techniques. These cutting-edge technologies allow structure determination of a wide range of proteins employing nuclear magnetic resonance, cryo-electron microscopy, Computational approaches such as molecular dynamics simulation and homology. SBDD may be broadly categorized into two parts: virtual screening and the De novo approach. The approach of De novo drug development takes advantage of three-D receptor data in order to discover tiny fragments which complement the biding site. In contrast, the virtual screening (VS) approach utilizes compound libraries already available to determine specific bioactivity hits.

It is a computational approach used to filter massive molecule datasets, and it has effectively been utilized to complement High Throughput Screening in Drug development. VS mainly aims to offer a cost-efficient and quick assessment of large databases of a virtual compound to screen efficient leads for synthesis and future research. In order to identify entities that are likely to bound a molecular target in interest, a screening using a virtual database can be done to filter huge compound libraries utilizing computational technology. VS primarily reduces the drug synthesis challenge to a massive extent by using enormous libraries of compounds that are pre-synthesized.

Now for the De novo approach, various techniques are:

8.3.1 Homology modeling

HOMOLOGY MODELING: another name for it is comparative modeling. It has focused on the notion that when two sequences share high identity/similarity, they will have a similar structure. Determination of target molecule structures follows specific drug target identification. A vast number of protein structure has not been discovered even after the technology advancement. In this situation, homology modeling can be helpful as it may be utilized to determine the protein structure using similar protein data.

With the following steps, protein 3-D structure can be determined using homology modeling (i) determining correct template, for query target sequence through BLAST search, (ii) alignment of sequence (iii) correcting alignments to assure the alignments of functionally or conserved critical residues (iv) generation of backbone (v) modeling of loop (vi) modeling of side chains through rotamer libraries (vii) model optimization through minimization of energy and (viii) model validation through stereochemical assessment utilizing residues in Ramachandran plot permitted regions (Fig. 8.2).

Structure predicted by homology modeling is assessed for its quality depending on the similarity degree between the template sequence and model. Query protein homology modeling will not produce a significant outcome if there is very low similarity. To overcome this obstacle, fold recognition may be used. Few server/methods for homology modeling are: (i) **Modeler**-it compares modeling between template sequences and given target to present protein model, for generating model non-hydrogen atoms are calculated by it. It can also be used for the optimization of protein and loop modeling. (ii) **I-TASSER**-it is an approach for hierarchical protein modeling, build on continuous execution of threading assembly refining program and alignment of secondary structure enhanced profile-profile threading (iii) **PRIMO** stands for Protein interactive modeling. It is a protein monomers homology modeling pipeline. It offers features that allow users to model ions and ligands in association with their target of protein. PyMod, Swiss Model, and MaxMod are some other recent servers/methods used for homology modeling.

The most precise computational approach for producing a credible structure model is homology modeling and is used consistently for various biomedical

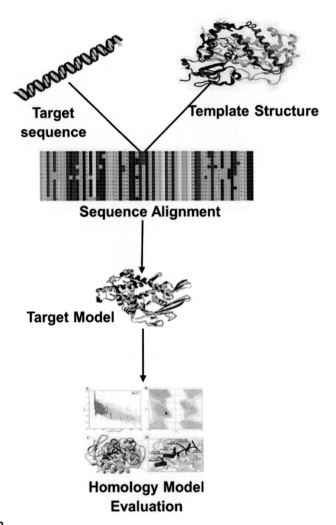

FIGURE 8.2

Homology modeling process.

applications. In general, computational work takes less than 2 h for a modeling project. Though it does not comprise, the need to visualize and interpret the results/ model may vary due to individual working experience and familiarity with protein structure.

8.3.2 Molecular docking

MOLECULAR DOCKING: It is a computational approach commonly utilized to quickly predict compounds' possible binding modes in target binding sites. On the basis of complementarity and conformation, it predicts small molecules' affinities toward their target molecule. In the field of drug development, this Insilco technique has attained a significant position. Over the last decades, it has evolved drastically and has become an essential component/tool for SBDD (Bender et al., 2021). It has also been known to be more effective/efficient than conventional drug designing approaches. The enormous increase in availability and accessibility to protein and small molecules databases and the advancement in computational technology has tremendously aided molecular docking. These computational advancement and biological targets 3D structure data availability are set to improve the efficiency and effectiveness of this technique. These developments will also accelerate its broad applicability in analyzing protein-ligand binding molecular interactions. Usually, docking of small molecules can occur in three ways: (i) through **RIGID DOCKING** in which ligand and target are viewed as rigid molecules; (ii) through **FLEXIBLE DOCKING**, where entities like target and ligand are regarded to be flexible; (iii) through **FLEXIBLE LIGAND DOCKING**, where the target is supposed to be stiff and ligand flexible. Molecular Dockings' main aim is to discover optimally binding ligands with receptor binding sites and identifying their strongest poses or binding orientation, which are energetically recommended. Binding Pose term can be defined as the ligand's confirmation/orientation relative to its receptor. It is either said to be a ligands conformation within target proteins binding site confirmed experimentally or as a hypothetical conformation modeled computationally. For identifying ligand-protein interactions, two critical elements required are the scoring function and search algorithm.

For identifying ligands' various conformations and poses for a given protein target, search algorithm is accountable. Whereas selecting the most desirable ligand/receptor binding modes or estimating generated poses binding affinities along with ranking them is done by a scoring function. An efficient and quick search algorithm is necessary, and for scoring function, it should be able to identify thermodynamics interactions and molecules' physio-chemical properties. A wide range of experiments are carried for finding ligands binding modes and selecting poses which are favored energetically the most. To achieve this purpose, tools for molecular docking facilitate sets of binding poses for various ligands and, as a usual scoring function, will be utilized to estimate generated poses binding affinities to choose the optimum binding mode. During recent years various tools or programs of molecular docking, including Dock, Surflex, Gold, and AutoDock, have been implemented and effectively employed in various drug discovery and research projects. The two essential components separating the diversity of available docking software are the scoring function and the sampling algorithm.

8.3.2.1 Sampling algorithm

In between two molecules, there is a large no of binding modes, and it would be time-consuming and expensive to produce all potential modes even with the help of modern computers' higher clock speed and advancement in parallel computing. Thus, the importance of algorithms was seen as they were required to filter out the significant conformations from unfruitful ones. In this regard, several algorithms were constructed, which can be categorized according to the no. of degrees of freedom they neglect. The simplest algorithm that was proposed evaluated molecules as two rigid entities, decreasing the degree of freedom to 6. DOCK program is an excellent illustration for this algorithm working. DOCK software was aimed to identify entities having a broad similarity or resemblance in shape to binding sites grooves/pockets. The algorithm works on taking a picture of the potential binding site available on the protein surface. This picture comprises different radii overlapped spheres which interact with just two spots on the macromolecule surface. Ligand's molecules are also seen as a sphere set that can almost pack the occupied space by ligand. The pairing rule is employed after the relevant ligand and protein representation as a sphere is finished. The principle of the rule is based on the fact that protein and ligand spheres can be paired with each other if ligand and protein spheres have similar internal distances, enabling user-specified tolerance. Hence permitting the program to determine spheres clusters which are geometrically similar on ligand and protein site. Other different software developed which uses the above matching algorithm (MA) consists of - LIDAEUS, SANDOCK, LibDock, etc. MA-based software has the speed advantage, but also the limitation of molecular flexibility lacks, etc.

Incremental construction is the following algorithm, which works on the idea of ligand being fragmented into different segments from rotatable bonds. Out of all, one specific segment is anchored to the surface of the receptor, and maximum interactions are shown by this anchor at the surface of the receptor being reasonably rigid. After setting up the anchor, every fragment is joined to each other gradually. The fragment having hydrogen bond interactions (accurate geometry prediction) have higher chances of getting first as they are responsible for ligand specificity. The algorithm becomes extremely robust and quick because, for the next iteration, most minor energy poses are selected. SLIDE, SKELGEN, and FLOG are some of the programs which use this algorithm. Limiting to medium size ligands is the only limitation this algo has.

Another algorithm in the section is the Genetic Algorithm which is utilized for determining the global minima. Darwin's Theory of Evolution is the inspiration for the algo. The genetic Algorithm includes ligands population filtered by scoring function, and every ligand is a potential hit. Crossover and mutation are used for the alteration of ligands population. New ligands are formed using a mutation operator from a single ligand, while information is exchanged between two or more ligands in the population using a crossover operator. It has been integrated into Autodock 4.0, GAMBLER, and DARWIN-like programs. Convergence Uncertainty is the limitation.

The hierarchical method is the approach in which ligands' low energy conformations are aligned and pre-computed. Pre-generated ligand conformations clusters are integrated into a hierarchy so that other similar conformations can be placed alongside one another in the hierarchy. In the end, this hierarchical information can be used by docking software after carrying out the ligand's translation, thus resulting in minimization of the outcome. E.g.: a clash between atom close to the ligands rigid center and protein in the given translation are rejected as conformations below in hierarchy will have descendants with the same clash. The software which uses this algo is GLIDE.

8.3.2.2 Scoring functions

The algorithm discussed above checks if sampling changes are performed accurately and rapidly to allow compound evaluation in the set computational period. These algorithms are then complemented by scoring functions. The ranking and evaluation of a predicted conformation of the ligand are essential. Scoring functions work in two scenarios; first, if you are interested in finding only how a biomolecule and ligand bind to each other, the scoring function determines the docked orientation that represents the intermolecular complex's true structure most accurately. Whereas in the other scenario, where you evaluate the various ligands, scoring functions determine not only the accurate docking pose but also rank ligands relative to each other.

Many simplifications and assumptions are used by scoring functions for estimating the possible complexes binding energy in a brief period. An adequate balance can be found between binding energy accurate estimation and computational time in cases of popular scoring functions. Over the past years, various scoring functions have been designed and can be categorized into three: Force field, empirical and knowledge base.

Force field scoring functions are formulated on physical atomic interactions like electrostatic and van der walls interactions, torsions, bond angles, and bond length. The parameters and functions of force fields are derived from both quantum mechanical calculations and experimental data. Some of the scoring functions included in this category are GoldScore, DockScore and HADDOCK Score, etc.

Empirical Scoring Functions: it is based on the fact that a sum of distinct uncorrelated terms can be used to approximate the binding energies of a complex. During binding energy calculations, various term coefficients are included, which are derived employing regression analysis utilizing binding energies which are experimentally obtained or perhaps from structural X-ray data. Compared to force fields, simpler energy terms are used in empirical functions, making it considerably more rapid in calculating binding scores. The first-ever software to use the empirical scoring function for binding free energy determination was LUDI. LigScore, SCORE2, HINT are other empirical scoring functions developed by including various empirical energy terms.

Knowledge Base Scoring Functions: These are acquired from structural data inherent in atomic structures predicted experimentally. Statistical analysis is used by the functions on complex structures to derive frequencies of interatomic contact between ligand and protein. The higher the interaction, the higher the occurrence frequency. MScore, BLEEP, DrugScore, etc., are some of the popular functions (Fig. 8.3).

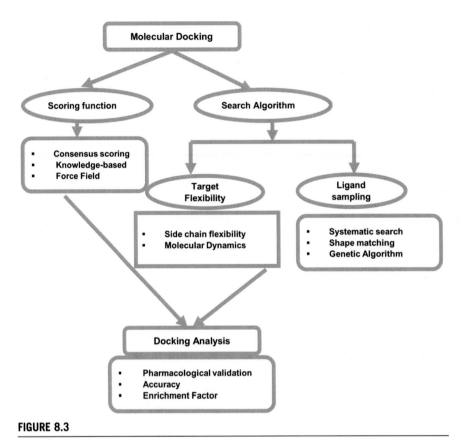

FIGURE 8.3

Flowchart of molecular docking.

8.3.3 Molecular simulation

MOLECULAR SIMULATION: Throughout mankind history, the research for novel medication has always been vital. The world population have been challenged consistently with various pandemics from 1800s, 1900s to the ongoing Covid 19 outbreaks, including potentially lethal diseases like cancer. Therefore, discovering drug being Scientifics community major concern. Remarkable collection of biological information is done on a daily basis recently, ranging from genetic sequence to 3D structures of protein and drug databases, providing good assistance for SBDD research. Over last decade, this has increased due to the fast development of quicker architectures and stronger computational techniques in an economical way. RCSB PDB is one of the uttermost important resource for MD simulations, and it offers 3D structural data which is experimentally determined (Hollingsworth and Dror, 2018). It is a global repository for managing and distributing macromolecules (nucleic acid and protein) structure data along with being a crucial asset for biomolecular modeling. It is a method/tool that calculates the system particle movement over a particular time period and then examines systems evolution. It is also stated as an approach used for evaluating motion and interaction of molecules. MS is among the most significant scientific and engineering discovery techniques that may be widely utilized in fields like material design and drug discovery. It now enables SBDD approaches which reflect the flexibility of the structure of drug target model system.

> *If most significant presumption is to be named, which motivates one to try to comprehend life, is that all entities are composed of atoms and all things done by living being is understandable by looking into atoms jiggling's and wiggling's.*
> **Richard Feynman (1965, Nobel Prize in Physics).**

Biophysicists have dedicated them for the more understanding of these jiggling and wiggling of the atom. Studying this jiggling and wiggling which is termed as molecular motion is very relevant to develop drugs. Ligands binding basic theory—"lock and key" where a motionless and frozen receptor was considered to incorporate tiny molecules without any conformational adjustment, is dropped in order to encourage binding models which account changes in conformations along with ligands and receptors random jiggling. General procedure for a molecular simulation starts with preparing or developing molecular system's computer model using homology modeling, crystallographic or NMR data. MD is a technique which works on the Newton's equations, (F = m*a). Then estimations of forces working on atoms in the systems is done. In short, forces originating from interactions among nonbonded and bonded atoms participate. Modeling of Dihedral angles is an approach which emulates differences in energy between staggered conformations and eclipsed using a sinusoidal function, whereas atomic angles and chemical bonds use simple virtual springs for modeling. These energy terms stated here are parameterized so that they can fit well in the experimental data and quantum-mechanical calculations. These parameters collectively are named "force fields" and depict different atomic forces contribution. These are utilized for evaluating forces among different atoms interacting and computing the system's total energy. CHARMM, GROMOS, and AMBER are few force fields commonly used in simulations. Based on their parameterization they differ principally yet similar results are generated by them. Secondly, newtons motion laws integration in the simulation creates a consecutive systems configuration which provide trajectories to determine particle velocity and position over time. Various attributes such as kinetic measure, macroscopic quantities and free energy can be calculated using these MD trajectories.

The CHARMM MD package is the oldest one developed by Martin Karplus. The GUI of CHARMM helps in preparing simulations input files by offering web-based graphics tool. Another package used for MD simulation which is relatively faster than the other packages is GROMACS (open -source) aka GROningen Machine for Chemical Simulations. It does not have force fields of its own like CHARMM and Amber have. Rather, force fields like GROMOS, Amber, OPLS and CHARMM can be imported in this package. Hence, using these packages, simulation process can be summarized as: (a) the system is initialized using zero total momentum,

(b) For each particle field forces are computed, (c) Newton's motion equation is integrated, (d) For a specified time, repeat b and c steps.

Even after being a competent technique, two major obstacles continue to restrict the MD simulation use: further refining of force fields are required and recurring simulations (>1 ms in length) are prohibited which leads to conformational states insufficient sampling in many cases. Even though few milliseconds' simulations are now feasible, obtaining sufficient statistics and thorough sample for conformational space are recommended to a number of alternative trajectories. For e.g., Of the above situation a simulation of a small system (approx. 25,000 atoms) for 1 microsecond, running on several processors requires multiple months for completion. In addition to the problems of high computational requirement of the simulations, force fields utilized are approximations of quantum-mechanical reality in atomic domain. Although many significant molecular movements may be predicted correctly by simulations, they are unsuitable for systems with significant quantum effects, such as transition metal atoms, engaged in binding. So, the technique is rigorous even when evaluating a single lead compound. Drug like compounds having extended unbinding kinetics have also been commonly found. The conventional MD approaches cannot still explain those sluggish unbinding events, even if operating on specialized hardware. This is the primary concern in rapid discovery of drug program, restricting the usage of MD simulations doing kinetic prediction. The sampling problem nevertheless has led to the creation of several new algorithms based on "enhanced sampling approaches." These accelerate the slow process description, enhancing the uncommon incidences defined by high free energy states (Table 8.1).

8.4 Ligand-based drug designing

LIGAND-BASED DRUG DESIGNING Presents an illuminating technique for the relation in-between physio-chemical and structural properties of ligands along with

Table 8.1 Difference between structure and ligand based drug designing.

S.No.	Structure-based drug designing	Ligand based drug designing
1.	Based on 3D structure of target protein obtained from NMR, X-ray crystallography	Based on known ligands of target protein
2.	Receptor structure is known	Receptor structure is not known
3.	Techniques involved- docking and molecular dynamics simulation	Techniques involved- QSAR, pharmacophore models, molecular similarity approaches
4.	Software used—AutoDock Vina, Glide, Autodock4.0, FlexX, Gold	Software used- MACCS-3D, ROCS, phase, Catalyst

their biological activities. It is also known as the Indirect Drug Design technique (Vázquez et al., 2020). This technique is implemented in the absence of target proteins 3D structural data. The data of ligands and their biological activity provided in this process is utilized to develop novel possible drug candidates. LBDD relies on the molecules' data bound to the active site of the biological target with their interest. These molecules are utilized for extracting an appropriate model that offers the major structural characteristics of a lead molecule that assists in the target molecule binding process. The target models are biologically active and are based on binding molecular information. This approach is made to discover new compounds that can interact with target molecule which are biologically active.

LBDD is frequently employed in pharmaceutical research. It works on the presumption that identical structural characterizes of compounds have similar biological activity and interacts with target molecules which are common. Molecule representation is the foundation of the LBDD approach. Numerical values for molecules physio-chemical and structural properties representation are called Molecular Descriptors. It is a highly cross-disciplinary field containing several theories. Ligand-based drug designing has the most popular approaches and can be classified into two parts: Pharmacophore modeling or Quantitative Structure-Activity Relationship (QSAR).

Methods of QSAR are founded on the assumption that biological activities are linked directly to biological activities, hence altering biological activities through structural or molecular variation. QSAR aims to be a procedure that constructs mathematical or computational models using chemometric techniques, which identifies a strong association between a set of functions and structures. The basic hypothesis for QSAR states that "similar activities are shown by compounds having similar physio-chemical and structural properties." Lead compounds producing expected biological activities are collected to make a library for finding the possible leads. A model is then created to determine the quantitative relationship between the biological activity of a compound and its physicochemical and structural characteristics. In order to quantitatively optimize the compound sets biological features, along with maximizing biological activities, a statistical model is constructed using the relations mentioned earlier.

8.4.1 Pharmacophore modeling

PHARMACOPHORE MODELING: Pharmacophore concept was first presented by Ehrlich around 1800. He described it as a "molecular framework that conveys (phoros) significant characteristics responsible for the biological activity of a drug." The underlying pharmacophore notion remained constant after a 100 years, but its deliberate meaning and scope for application were substantially broadened (Kaserer et al., 2015). IUPAC's recent definition of a pharmacophore model" is a group of electronic and steric properties essential for ensuring the effective supramolecular interactions and triggering biological response with a particular biological goal. The concept of Pharmacophore modeling pre-exists than any electronic computer, yet it works as an effective tool in CADD. Molecule units like atoms or groups with specified characteristics associated with molecular detection can be turned into a pharmacophore feature. These molecular patterns are characterized as anionic, cationic, hydrophobic, aromatic, H-bond acceptors and donors, or any potential combination. Various compounds may be compared at the pharmacophore level, commonly termed "pharmacophore fingerprinting." The pharmacophore can be termed a "query" if only a few pharmacophore characteristics are considered a 3D model. There are some properties grouped in a specific 3D layout in a pharmacophore model. Every characteristic is generally described as a sphere. The characteristics may be designated as an individual characteristic or any combination logic gates: "NOT," "OR," and "AND." Typically, certain pharmacophore properties are employed for screening compounds' small-molecule libraries. All compounds exist as their low−energy bio-relevant conformations in these libraries. Every conformation in these libraries is suited to pharmacophore query through alignment with molecules pharmacophore characteristics, hence composing the query. A molecule is considered a hit molecule if it can be accommodated in the spheres representing query characteristics. Sometimes pharmacophore query may become extremely complex in finding hit molecule from a particular library, and then the partial match is permitted. In these circumstances, just particular qualities are matched, which are regarded vital to activity. Diverse ways are available to design pharmacophore models either manually or via automated algorithms based on the scenario and type of experiment.

Here we are going to discuss automated algorithms. Generally, the pharmacophore generation involves two steps: construction of ligand conformational space in training set for representing ligands conformational flexibility, ligand alignment of many ligands, and determining critical common chemical elements to build models for pharmacophores. The main problem in pharmacophore modeling and the significant methodologies are represented by conducting molecular alignment and management of conformational ligand flexibility. Pharmacophore modeling-based software's which are available are—DISCO, PHASE, MOE, and HipHop, etc. All these software differentiate only on the basis of algorithms they are using to handle molecule alignment and ligands flexibility.

Even after significant advancement, there are still major challenges in pharmacophore modeling that exist. Ligand flexibility is the first challenge for modeling. There have been two strategies to fix the problem: the pre-enumerating method is the first where multiple conformations are pre-computed and stored in a database. This method has a low cost for computing as an advantage for carrying out molecular alignment at the cost of the need for a potential storage capacity. For the second method, analysis of conformations is done, known as the on-the-fly method. It requires no bulk storage but may require more time on the CPU for performing a rigorous optimization task. Pre-enumerating approach has been shown to exceed the on-the-fly calculation method.

The second issue in pharmacophore modeling is molecular alignment. According to the fundamental nature of alignment methods, they can be grouped into two sets: property and point-based method. The points may be further divided as chemical features, fragments, and atoms in a point-based approach. The most significant constraint related to this technique is the requirement for pre-set anchor points, as constructing these points can be difficult in different ligands (Fig. 8.4).

Nevertheless, another obstacle is the experimental aspect of correctly choosing training set compounds. This issue which seems non-technical and easy, usually puzzles users, including experienced ones. The ultimate pharmacophore model generated is affected by the dataset size, ligand molecule type, and diversity of chemistry.

Although the pharmacophore notion has limits, various remedies may be used to conquer them at any moment. In light of this adaptability, pharmacophore modeling is predicted to continue to be a significant part of CADD and has so much scope, advantages, and prospects shortly.

8.5 ADMET

Drug development's fundamental objective is to obtain a compound that shows therapeutic effect into a medicine form, which can be used for patient's dose (Van de Waterbeemd and Gifford, 2003). The drug is required to be reached at the site of the problem, employ its pharmacological effects, and get disposed of promptly. The characterization of ADME features helps analyze and describe how pharmacokinetics processes occur to give a novel medicine with safety considerations on which risk-based evaluations may be carried out. ADMET is the method that explains the pharmacokinetics disposal of the drug, or activities are done to a drug by the human body. In the pipeline of drug development, ADMET data can be extracted from several stages. Drug developers may apply chemical changes to drug candidates to maximize ADME qualities in the discovery and optimization process. In vivo and in vitro studies provide crucial data to fulfill regulatory requirements, from Drug progression via preclinical development and clinical phases for pharmacists to make reasoned judgments.

With the development of novel compounds, the requirement for ADMET Information begins. Moreover, this information can influence the choice of proceeding with the synthesis via combinatorial chemistry techniques or with traditional chemistry methods. Computer techniques are the only choice for this information, although prediction at this time is not 100% accurate but still acceptable. More robust mechanical models are necessary when a molecule series is concentrated on a lead and further streamlined to a clinical candidate. To develop in-silico models that enable a rapid assessment of various ADMET features, a comprehensive understanding of the links between molecular structure and properties with ADME parameters is employed. In addition to this information, other properties can also be predicted that offer data on frequency and size of the dose, like bioavailability, distribution volume, oral absorption, and brain penetration. The available experimental data in academia has led to significant efforts in developing models for the prediction of ADME-related physical-chemical parameters, such as lipophilicity. Even

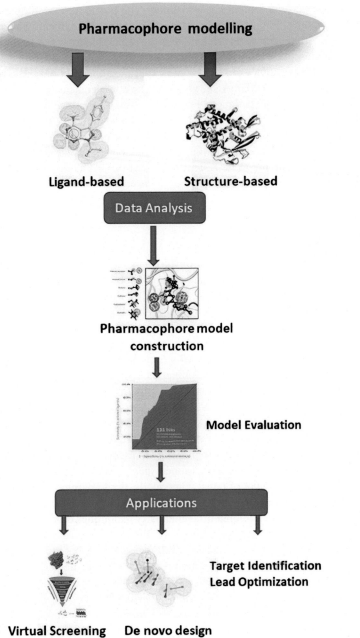

FIGURE 8.4

Pharmacophore modeling.

after the relevance, the determination of pharmacokinetic features such as distributing volume, clearance, and half-life directly from the molecular structure progresses less slowly because of the absence of data published. The estimation of many metabolic and toxicity aspects is also under-developed.

To understand what computational tools are required by ADMET, we need to consider two aspects: molecular modeling and data modeling consisting of a distinct toolbox. QSAR methodology is usually used for data modeling. Since the 1960s, QSAR and QSPR (quantitative structure-property relationship) research has evolved, involving various physio-chemical and biological data. The research done by these employs statistical tools for finding associations between provided features and a range of structural and molecular descriptors of a given molecule. During the past 40 years, a broad range of descriptors was produced for use in QSAR analysis. The descriptors subset might be beneficial in determining ADME Properties. The descriptors include hydrogen bonding and molecular size, while other descriptors can be quantum-chemical concepts or are merely topological.

In molecular modeling: techniques like protein modeling are involved, quantum mechanical methodologies are employed for evaluating the possible interaction between proteins (like p40s) intricated in the ADME process and the small molecules. For these proteins, 3D structural data is required to be developed by homology modeling of the associated structures if the structure of a human protein is not obtainable. An additional technique to analyze the possibility of a small molecule interacting with a particular protein when there is no structural protein knowledge is to employ PHARMACOPHORE models constructed on the superposition of known protein substrate.

For ADMET parameters, finding good predictive models usually depends on selecting suitable molecular descriptors, appropriate mathematical strategies, and a suitably huge pool of experimental information for model validation. There is increasing understanding of which of the descriptors and QSAR techniques available are most suited, though alternative solutions with the exact prediction capacity frequently seem to exist. It is crucial to study, in particular, how training set size impacts the model choice.

8.5.1 Adsorption

For a substance to access tissue, it usually has to be carried in the blood circulation—typically via mucous surfaces in the digestive system, prior to being absorbed by the target cell. Hence, it is the mechanism through which the drug reaches the bloodstream. Although there are numerous different pathways, the two most popular modes of administration are oral and intravenous. The advantage of giving drugs intravenously is that the drug skips the absorption phase and directly enters circulation. Nevertheless, a lot of medication is prescribed to be taken orally as it allows the patients to administer it themselves. During the ingestion of Xenobiotic, it passes via the gastrointestinal tract, and then through the portal circulation, it moves to the liver, entering systemic circulation, which makes the drug reach its active site.

More minor compounds often cross membranes in this procedure, sometimes by passive transport, but commonly through proteins called drug transporters. In several phases of the pharmacokinetic trip, drug transport might be the primary element of a drug disposition, and preclinical research must be carried out to offer data on drug interaction with different transporters—like inhibitors or substrates.

There are four critical routes of administration:

(a) Injection directly to the bloodstream
(b) Dermal Application
(c) Inhalation
(d) Ingestion via the digestive tract

The drugs have to penetrate through a membrane prior to entering the bloodstream when administered via dermal contact, inhalation, or ingestion. While in the case of injection, drugs directly enter the bloodstream. There are specific ways, four in common, by which a substance can pass a membrane and enter circulation: (a) **Active Diffusion**: the drug molecules in ATP form need the energy to cross the membrane. It is an energy-dependent mechanism. (b) **Passive Diffusion**: the standard uttermost method by which drug is absorbed. In this, drug molecules travel from a high concentration area to a low concentration area. (c) **Endocytosis:** when a large drug is transported across the membrane through membrane invagination. (d) Facilitated Diffusion: drug molecule traveling from high concentration area to low concentration area using carrier proteins in the membrane. Bioavailability is influenced by the method of delivery, a measure of how many drugs is taken in their unmodified form. Bioavailability can be found by evaluating the concentration of drug plasma over time. The only way to achieve bioavailability by 100% is by intravenous injection. Reduced availability will be observed in case of other ways of drug administration. Not every drug molecule will reach the bloodstream. For example, the drug that is initially consumed goes under metabolism, which excretes some drug molecules before entering the circulation of blood. Many parameters, including solubility, molecular weight, ionization, the topological polar surface area, and other physicochemical features, might influence the absorption of drugs. For evaluating the potential of drug quantity reaching circulation after oral consumption, absorption data can be beneficial. After oral absorption, the first-pass impact (with other components) determines bioavailability.

8.5.2 Distribution

When the drug gets absorbed, it goes from the absorbing location to various body tissues like organs, muscles, and generally to various extents. The distribution is usually achieved via the circulation from one body region to another; however, it can also happen from cell to cell. The drug compound is submitted to several distribution procedures which serve to diminish its plasma concentration after entering into the systemic circulation, whether by absorption from any one of the several extracellular locations or by intravascular injection. To measure efficacy, the researchers assess

the rates at which the chemical reaches various places and the extent of the dispersion. Certain drugs are easy to transfer, while others are not. The reversible movement of medicine from one region to another can be defined as Distribution.

Polarity, molecule size, regional flow of blood, and serum protein bonding, producing a complex, are some elements impacting drug distribution. Some natural bar-
riers, such as the blood-brain barrier, can provide a significant concern. Different
in vitro research might assist in compiling more information about the distribution
of a substance. E.g., Studies of drug transmitters assist in identifying proteins that
are responsible for driving drugs into and out of cells, and permeability testing
can characterize the compound's potential for entry into cells.

8.5.3 Metabolism

Metabolism of drugs can be defined as the biotransformation of a drug into hydrophilic metabolites primarily through tissues and organs like skin, liver, digestive
tract, or kidney so drug molecules can be extracted from the body through excretion.
Most of the drug metabolism of a small molecule is performed by the redox enzyme
(cytochrome p450) in the liver. The initial chemical becomes a new substance
termed metabolite when metabolism takes place. It disables the provided dose of
the parent medication when metabolites are pharmacologically inactive and thus
generally lessens the body's impact. Drug metabolism includes enzymes and
numerous investigations that may be necessary for the identification of key metabolites and related metabolic pathways. Chemical metabolism can lead to toxicity, for
example, through the creation of toxic by-products or metabolites. Adverse
Outcome Pathway (AOP) is something researchers draw a drug candidate's specific
metabolic pathways. It offers information required for determining the drug's potential toxicity and safety.

For meeting regulatory submission expectations and validating important actors
in the metabolism of drugs, specific drug metabolism research has been carried out.
The research includes characterization of metabolites, metabolic stability, and identifying metabolite for the elucidation of metabolites among species and establish
whether any of these is unique to humans or disproportionately greater in humans
compared with preclinical species.

8.5.4 Excretion

The process through which the elimination of metabolized drug compounds occurs
in the body is called excretion, generally done through the kidney. Usually, all metabolites and material related to drugs like parent drugs are ultimately removed from
the body. Characterization of the excretion routes is fundamental; it mainly occurs
from the liver and kidney, as mentioned earlier; however, tears, sweat, and breath
can also be used as excretion doors. Normal metabolism can be impacted adversely
by the foreign substance's accumulation unless excretion is completed. Scientists are
currently working on finding out about the pathways taken by the drugs to leave the

body and how promptly the drugs can be excreted while keeping in mind that excretion pathways can be influenced by molecular charge and size.

It is also stated that every drug is not entirely excreted out; adverse effects can happen when metabolite or chemical by-products accumulate. Research in "in vivo excretion" can be helpful in both identifying compounds excretion routes and characterize material clearance related to the drug while surveilling exposure of metabolite and drug in plasma and different compartments. Radiolabeled molecules are used in animal mass balance research for characterizing the excretion rate and path of the drug. This research showed the whole scenario of what rated drug is excreted from the body along with the quantitative evaluation of feces and urine. Other supporting research may provide further data to study lymphatic partitioning rate, excretion via milk, biliary excretion, and more.

8.5.5 Toxicity

Toxicity accounts for several compounds not making to the market and removing a considerable number of compounds from the industry that were once approved. Approx. 20–40% of research drug development failures have been estimated to be due to toxicity concerns. In silico technologies, which are commercially accessible to predict possible toxicity, may be grouped into two classes. In the first technique, expert systems are involved, which generate modeling based on codifying and abstracting knowledge from scientific literature and human experts. The second strategy is primarily based on the production of chemical structure descriptors and statistical analyses of correlation between the toxicological endpoint and the descriptors. The crucial aspect of any in silico technique is the data quality used for the purpose of model development. The restricted accessibility of toxicity data for the public domain has constrained the toxicology endpoints quantity foreseen by the commercially accessible system. Mutagenicity and Carcinogenicity are the central objectives of modern software packages. Although some programs also incorporate knowledge bases and models for additional objectives, including sensitization, irritation, neurotoxicity, immunotoxicology, and teratogenicity.

8.6 Drug repurposing

It is a strategic approach for identifying novel uses for authorized or researched medicinal products outside the initial medical indication scope (Pushpakom et al., 2018). It has various other names like Drug profiling, drug recycling, drug retasking, therapeutic switching, and drug rescuing. This entails creating novel therapeutic applications for drugs already available in the market, including discontinued, experimental, abandoned, and approved drugs. The traditional method of determining drugs is a laborious, risky, costly, and time-consuming task.

The new drug repurposing technique can probably be used over the traditional drug discovery program by reducing greater chance of failure, longer development periods, and high monetary costs. This technique offers many benefits over the development of an entirely novel medicine for a particular indication. First, and probably most crucial, the chance of failure is decreased as the repurposed drug has previously been demonstrated to be adequately safe in preclinical models. It is less likely to fail from the safety viewpoint in later effectiveness studies, at least if early-stage studies have been completed. Secondly, it is possible to minimize the time frames for developing the drug, as principal safety evaluation, preclinical testing, and in few situations, creations of the formulation have already been done. Thirdly, there is a requirement of less investment; however, this will rely mainly on repurposing candidates' development process and stage. In conjunction with these benefits, there is the prospect that investment in the repurposed drug development has the potential of resulting in faster and less dangerous approach. Ultimately drug recycling might disclose novel pathways and targets which can be exploited further.

Drug repurposing first example that happened in the 1920s by accidental discovery. Many strategies have developed since a century of advancement, which has accelerated the drug repositioning process. A Few examples of the best drugs which are the results of DR are aspirin, sildenafil, methotrexate, valproic acid, and minoxidil. Sildenafil approach in DR meant that it was primitively developed for hypertension treatment; however, currently, it is also used for treating erectile dysfunction.

On-target and Off-target are two strategies for Drug repurposing. Drug molecules, known pharmacological methodology, is put on a novel therapeutic indication in on target strategy. In this strategy, drug molecules' biological target is similar; however, there is a difference in disease. The strategy can be demonstrated using minoxidil example: it works on similar targets but generates two distinct therapeutic effects. Being an antihypertensive vasodilator, it permits more nutrients, blood, and oxygen in the follicles of hair, hence helping in treating androgenic alopecia (male baldness). In the case of the Off-target strategy, there is no knowledge of pharmacological mechanism. For novel therapeutic indications, drug candidates and drugs work on novel targets. Hence both indications and targets are new. An example of an off-target strategy is aspirin; it has been used in treating various inflammatory disorders and pain. It functions as a blood coagulation suppressor, hence employed in strokes and heart attack treatments.

Drug repurposing also has two approaches: in silico based and experiment-based approaches. Activity-based repurposing is another name for the experimental-based approach; it applies to the initial drug screening for novel pharmacological indications on the basis of experimental studies. Various other approaches included in this are cell assay, clinical, target screening, and animal model approach. It needs no structural data of the target protein. Though It has a lower false-positive rate in screening but it is laborious and time-consuming.

Whereas in silico repurposing virtually screens large chemical/drug libraries of public databases utilizing cheminformatics/bioinformatics tools and computational analysis. Potential bioactive molecules are identified based on molecular interaction between the drug molecular and the target protein. This approach has the advantage

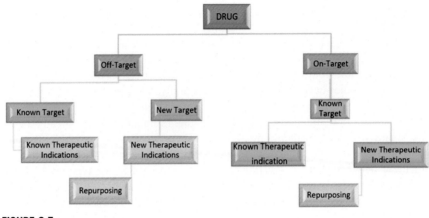

FIGURE 8.5

Drug repurposing flowchart.

of It being labor and time-efficient. However, it has higher chances of false-positive results during screening (Fig. 8.5).

Drug repurposing methodologies can be classified into three major categories according to the quality and quantity of biological, toxicological, and pharmacological activity data. These include: (a) diseases oriented, (b) target-oriented (c) drug-oriented. In disease-based methodology, drug repurposing is significant when more disease model data is accessible. Drug repositioning can be led by treatment or disease on the basis of data accessibility provided by genomics, phenotypic data, metabolomics, and proteomics data related to the disease process. Hence, there is a requirement of building a particular disease network, identification of protein molecules causing diseases, and genetic expression recognition.

The target-based methodology involves virtual high throughput and in silico screening of compounds or drugs from drug databases like molecular docking of the drug, followed by high throughput screening in accordance with protein biomarker or molecule of interest. This methodology has a considerable success rate for discovering drugs as disease mechanisms and pathways are represented by most of the biological targets.

Biological activities, toxicities, adverse effects, and drug molecule structure features are measured in drug-oriented methodology. It is based on the identification of molecules having biological effects on animal assays. The drug-oriented methodology has drug discovery and traditional pharmacology principles as its basis, where research is generally done to determine drug molecules' biological efficacy with unknown biological targets.

DR provides many pharmaceutical businesses the option to generate lower investment pharmaceutical products. Its mixed strategy offers more effective and speedy chances for determining repositioning drugs. From a commercial perspective, many diseases require treatment from novel medication with possible economic consequences and market demand. For instance, medicines for uncommon/ neglected diseases can be discovered with a vast potential market. Consequently, there is a chance to repurpose medications to cure uncommon, neglected, or orphan diseases or diseases difficult to cure.

References

Batool, M., Ahmad, B., Choi, S., 2019. A structure-based drug discovery paradigm. Int. J. Mol. Sci. 20. https://doi.org/10.3390/IJMS20112783.

Bender, B.J., Gahbauer, S., Luttens, A., Lyu, J., Webb, C.M., Stein, R.M., Fink, E.A., Balius, T.E., Carlsson, J., Irwin, J.J., Shoichet, B.K., 2021. A practical guide to large-scale docking. Nat. Protoc. 16 (10), 4799—4832. https://doi.org/10.1038/s41596-021-00597-z.

Hollingsworth, S.A., Dror, R.O., 2018. Molecular dynamics simulation for all. Neuron 99, 1129. https://doi.org/10.1016/J.NEURON.2018.08.011.

Kaserer, T., Beck, K.R., Akram, M., Odermatt, A., Schuster, D., Willett, P., 2015. Pharmacophore models and pharmacophore-based virtual screening: concepts and applications exemplified on hydroxysteroid dehydrogenases. Molecules 20, 22799. https://doi.org/10.3390/MOLECULES201219880.

Ou-Yang, S.S., Lu, J.Y., Kong, X.Q., Liang, Z.J., Luo, C., Jiang, H., 2012. Computational drug discovery. Acta Pharmacol. Sin. 33 (9), 1131—1140. https://doi.org/10.1038/aps.2012.109.

Pushpakom, S., Iorio, F., Eyers, P.A., Escott, K.J., Hopper, S., Wells, A., Doig, A., Guilliams, T., Latimer, J., McNamee, C., Norris, A., Sanseau, P., Cavalla, D., Pirmohamed, M., 2018. Drug repurposing: progress, challenges and recommendations. Nat. Rev. Drug Discov. 18 (1), 41—58. https://doi.org/10.1038/nrd.2018.168.

Song, C.M., Lim, S.J., Tong, J.C., 2009. Recent advances in computer-aided drug design. Briefings Bioinf. 10, 579—591. https://doi.org/10.1093/BIB/BBP023.

Vázquez, J., López, M., Gibert, E., Herrero, E., Javier Luque, F., 2020. Merging ligand-based and structure-based methods in drug discovery: an overview of combined virtual screening approaches. Molecules 25, 4723. https://doi.org/10.3390/MOLECULES25204723.

van de Waterbeemd, H., Gifford, E., 2003. ADMET in silico modelling: towards prediction paradise? Nat. Rev. Drug Discov. 2 (3), 192—204. https://doi.org/10.1038/nrd1032.

A machine learning approach to bioinformatics

9.1 Introduction to machine learning?

A machine learning algorithm is a statistical computation method used in software to detect hidden patterns that are not obvious in a dataset and make reliable statistical predictions of similar new data. Machine learning techniques attempt to find a pattern in a particular dataset; using these learned patterns, a similar pattern in a new dataset is identified. Machine learning processes are somewhat close to statistical modeling and data collection. They look into the data to find trends and accordingly learn the pattern or parameters. We usually know about machine learning through social media connection suggestions, online shopping product recommendations, spam filters in our email inboxes etc. In the last 30 years, there has been a surge in the use of machine learning techniques to solve biological problems, with successes in gene prediction, protein function prediction, cell image recognition, pathway analysis, protein structure prediction, drug molecule and toxicity prediction, and so on. To solve a problem using a machine learning algorithm, we require data about the concerned problem, which consists of features or characteristics of problems. For example, the concept of automatic annotating proteins' function has become more prevalent in recent years because more and more proteins are being discovered and identified. In an attempt to comprehend the molecular mechanism of biology, it is important to assign the function to large-scale proteins. But only a tiny percentage of the more than 179 million UniProtKB proteins have experimentally supported gene ontology (GO) annotations. Here, machine learning can be applied as we have annotated proteins fed as examples to an ML algorithm. The ML algorithm can learn patterns from examples and help us predict newly discovered proteins' functions. Features are used to show examples of an ML algorithm. For the present case, features may be protein sequence, Position-Specific Scoring Matrix (PSSM), representing conservation and homology, protein—protein interaction (PPI) network data, motifs, domains etc. In simple terms, features are the information about a particular example under observation (Larrañaga et al., 2006).

Choosing descriptive, discriminating, and autonomous features is an essential step in training efficient machine learning algorithms. After feature selection, the next step is to identify the problem's category, based on which we choose an ML algorithm. We will discuss the different types of machine learning algorithms in

All About Bioinformatics. https://doi.org/10.1016/B978-0-443-15250-4.00010-1

the subsequent sections. Then we train the models and evaluate their performance. According to recent machine learning footprints in life science, it seems to be helping researchers to address a variety of tiresome issues to gain greater understanding and opportunities for a prosperous future (Greener et al., 2021).

9.2 Types of machine learning systems

Machine learning has evolved from a science fiction fantasy to a widely used method in our community. It has such a significant impact on the outcome that incorporating machine learning algorithms into the process has become a deciding factor between success and failure. These algorithms are challenging to apply, and considerable effort has been expended in this area to get it to where it is today. It is important to comprehend what we want our algorithms to accomplish and the benefits they provide. Certain problems, such as predicting temperature, pH, or pressure, require a continuous numerical value as an output referred to as regression problems. In comparison, some questions require categorization, such as whether they are positive or negative or warm or cold. Additionally, there are instances where we expect machine learning models to classify examples based on the similarity and dissimilarity of their features. Machine learning models are generally classified into three categories based on their outcomes: supervised learning, unsupervised learning, and reinforcement learning.

9.2.1 Supervised learning

The most frequently used method of machine learning is supervised learning. It's the simplest concept to understand and put into practice. It's similar to using flash cards to teach a child something. We can feed a supervised algorithm data sequentially in the form of example-label pairs, allowing the algorithm to evaluate the label for each example and provide feedback on whether the label was estimated correctly. Finally, the algorithm will discover how to precisely interpret the relationship between instances and labels. Once thoroughly trained, a supervised learning algorithm will be capable of observing a new sample that has never been seen before and evaluating a suitable label for it. As a result, supervised learning is frequently referred to as task-based. It focuses on a single problem, training the algorithm with ample of examples until the task can be completed successfully. Additionally, supervision learning can be classified as regression or classification. Classification is generally used to predict a label, whereas regression is usually used to predict a quantity.

Regression is a technique for developing models that predict continuous values based on their input variables. In regression problems, the mathematical mapping function (f) from the input variable (x) to the output variable (y) is determined. Assume we have a dataset containing the mature height, weight, diet, and gender of our parents. Because height or weight is a continuous variable, training a supervised algorithm on these features to predict the child's height or weight is referred to as a regression problem. Classification, on the other hand, is a technique for identifying a model that divides input data into a number of distinct classes or labels. In light of the aforementioned data set of height and weight, the classification task would be to use the adult height, weight, diet, and gender of the parents to predict the child's height as "Above Average" or "Below Average."

9.2.2 The below are the most commonly used supervised algorithms

9.2.2.1 Linear regression

Linear regression is a supervised learning algorithm used for the prediction of continuous variables like sales, age, product price, salary and future weather etc. Linear regression is represented by a linear equation in which one variable is independent (x), and the second one is the dependent variable, or the output (y).

If the input and output variables are linearly related, a line could be drawn with the data sets to depict the relationship between the variables, as shown in Fig. 9.1. The equation for a basic linear regression model with a single independent variable x and a single dependent variable y will be:

$$y = mx + b$$

It is the equation of the line, where "x" is the input variable, y is the output variable, "m" is the slope, and "b" is the y-intercept. Linear regression aims to find the best values for "m" and "b" based on "x" and "y" data, so that the average difference

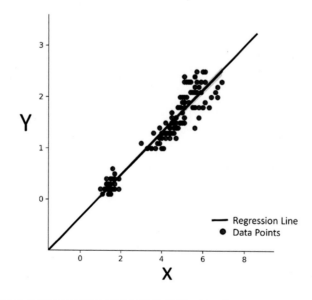

FIG. 9.1

Linear regression: Data points and the estimated best-fit regression line.

between the data points and the line is as small as possible. The difference between the data points and the line is referred to as an error of the model or the line, and each data point has its own error. The value obtained as these errors are squared and added is known as sum squared error, and the average of this sum is known as mean square error. The total error between all data points and a line is referred to as the mean square error, which states the model's efficiency.

Since "m" and "b" may have a wide range of values, regression iteratively finds the best line by fitting several lines and selecting the one with the lowest mean squared error (Fig. 9.2B). If there's more than two variables in a problem, the same approach is used; this is multivariant linear regression.

When there are more than two variables or more dimensions in the data, the regression line is a plane (in three dimensions) or hyper-plane (in more than three dimensions), and the equation is:

$$y = m1x1 + m1x1 + m2x2 + m3x3 + \dots \dots mnxn + b$$

Here, the dependent variable is y, and the independent variables or features are x. The coefficients of the features m1, m2, m3, m4, and mn provide an understanding of each features's contribution in the measurement of "y". Finally, the letter "b" is a bias term. We've seen how linear regression can be used to model a continuous variable, but the coefficients can often provide valuable knowledge about the significance or function of each feature in the estimation of y. We may also deduce approximate linear equations for problems, giving us an approximate mathematical model of the problem (Altman and Krzywinski, 2015).

9.2.3 Logistic regression

Classification problems involve recognizing spam or ham messages, deciding if a tumor is benign or malignant, determining whether or not a person has a disease,

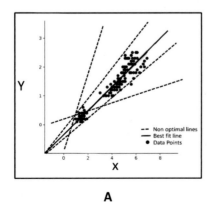
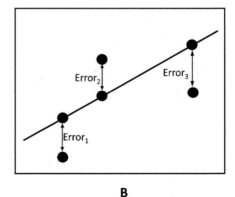

A   **B**

FIG. 9.2

(A) Best fit line among the various non-optimal lines and (B) Distance (errors) between the data points and the line.

and so on. These are regarded as binary classifications since there are just two classes. In general, two classes are indicated by the letters' "0" and "1." For example, a benign tumor is can be represented as "0," whereas a malignant tumor can be represented as "1." As a consequence, the model has to predict the output values of "0" or "1." We cannot use linear regression equations since the output varies from −(ve) infinity to +(ve) infinity (Lever et al., 2016). So, for classification, we have a sigmoid function, which has an "S" shaped curve, as seen in Fig. 9.3.

9.2.4 K-nearest neighbor

The KNN algorithm can be best illustrated by the expression "A man is known by the company he keeps." When a new disease is discovered, the doctor will recall previous patients who had the same symptoms. As a result, symptoms can be thought of as features, and labels as diagnostics or treatments for the disease (Parry et al., 2010). In KNN, the sorted training data set serves as a model representation, and no additional learning is required. Fig. 9.4 depicts the data points for two groups, blue and green, respectively. If we use K = 3 to create a new prediction event, the model will look for the three closest data points before assigning the new instance to the majority class with the highest probability. The diagram shows that the K value is an important factor in determining the model's behavior. Low K values reduce model accuracy because noise has a greater impact on it, whereas high K values make the KNN algorithm computationally intensive. As a result, for greater accuracy, an optimal value of k is chosen. It is critical to select the appropriate k value to balance accuracy and computing power needs.

Various distance measures are used to determine which of the nearest neighbors is the closest. The most commonly used distance calculation is the Euclidian

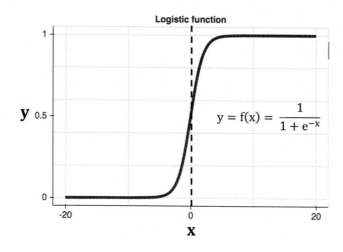

$$y = f(x) = \frac{1}{1 + e^{-x}}$$

FIG. 9.3

The logistic or sigmoid function.

K Nearest Neighbor

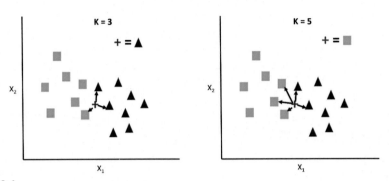

FIG. 9.4

Sinigicae of "K" in K-nearest neighbor algorithm.

distance. The Hamming distance is also used, which is the sum of the absolute differences between unit vectors. The dimensionality issue arises as the number of features increases. The problem in a high-dimensional space is that all points are far away from one another or their neighbors (Fig. 9.4).

Several advantages of using KNNs include the following: The algorithm is straightforward and simple to train. Due to its adaptability, the KNN can be used for both classification and regression. KNN calculates the average value of the k nearest data points in regression; it is more consistent and reliable when dealing with a large training set. A few drawbacks of KNNs include the algorithm's requirement for an exponential increase in computing power as the value of k increases. Its computational cost for predicting new data is quite high. When it comes to categorical features, KNN fails miserably.

9.2.5 Decision trees

The decision tree uses a tree-like hierarchical decision-making model to learn basic decision-making rules and predict the target variable's value. It is a general technique that has applications in a number of real-world contexts, including civil engineering, law and business, and is also widely used in machine learning. Decision trees are used to accomplish specific goals, such as a flow chart or a series of decisions (Kotsiantis, 2011). Fig. 9.5 shows the standard structure and terminology of the decision tree. Each node is a test or condition, and the branches are the outcomes of such tests or conditions (Fig. 9.5).

- Root Node: This is where the entire population or sample is split into two or more segments.
- Branches: Branches are the result of node choices.

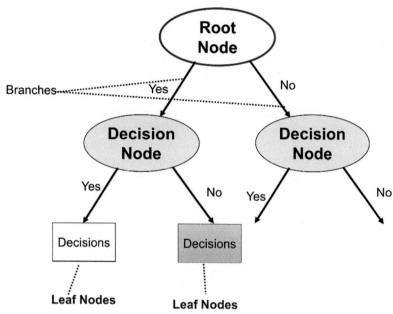

FIG. 9.5

Standard structure and terminology of the decision tree.

- Decision Node: When a sub-node is divided into additional sub-nodes, the decision node is formed.
- Leaf Node: Leaf nodes, also known as terminal nodes, are the final outcomes of all chain decisions.

Splitting is the method of separating a node into a number of sub-nodes depending on the test condition. Decision trees establish test conditions based on the purity or impurity gain form the decision in the lower subsets. The depth of the decision tree is increased with the number of decision nodes lead to the over-fitting of the model. As a result, we must restrict the creation of decision nodes or boundaries in real-world data while at the same time tolerating minor impurities in leaf nodes. The technique of deciding the depth of three is called tree pruning.

The following are some of the advantages of decision trees over other supervised learning algorithms:

1. Visually depicting and easy to understand
2. There is very little or no data processing needed.
3. It can be used for numerical as well as categorical results.

Decision trees, like every other machine learning algorithm, have their own set of drawbacks, including:

1. Excessive parameter tuning is required to avoid overfitting. Methods such as pruning and the determination of the minimum number of samples are used to solve the problem of overfitting.
2. The resulting tree may not be a globally optimal decision tree, because most decisions are made at individual nodes.
3. Most of the time a single tree is not enough to yield successful results.

To resolve these drawbacks of decision trees, multiple decision trees' ability is used. A random forest is a collection of decision trees, as the name implies. A random forest classifier trains multiple decision trees on small samples of the training data points. These small samples are selected on a random basis with substitutes, resulting in a large number of decision trees. Bootstrap sampling is a tool for the selection of small samples. In order to produce the final result, the random forest algorithm integrates the contribution of individual decision trees. Since it is a set of multiple decision trees, the model is also known as the ensemble type model, where two or more models are used for deducing results. The result of the forecast depends on the overall observation of all trees, i.e., the majority vote. Random forests are much more reliable and resilient than single tree species. Random forests are more accurate than decision trees for two major reasons:

1. The random forests, unlike the decision trees, are not pruned, so the characteristic space is divided into smaller and smaller areas.
2. Each random forest tree learns from a random array, and each node divides on the basis of a random set of characteristics resulting in tree diversity.

9.2.6 Support vector machines

The support vector machine are generally used as supervised learning algorithm in machine learning. In addition to classification they can also be used in regression problems. In this section we will discuss the application of SVMs for the classification task. Support vector machines are linear classifiers, because they make linear decision boundaries. It aims to find a hyperplane or decision boundary that is the best fit and separates n-dimensional space into separate classes or groups (Wang and Lin, 2014). When we place the new instance to a trained SVM, it classifies the new data point into one of the categories. The optimal decision boundary is known as a hyperplane. The hyperplane is a flat decision boundary with dimensions N-1 for an N-dimensional dataset. For 2D data, visually, it will be a line, and for 3D data, it will be a plane separating two groups. For applying support vector machines, the data should be linearly classifiable, which means a line or plane should exist form where the groups or classes can be separated. The objective of SVM is to find that line of the hyperplane.

In Fig. 9.6A we can see that the dataset is linearly separable; however, there can be many lines that can separate the data into different classes. Support Vector Machine selects the optimal points which help in creating the hyperplane. The optimal points are the data points that are very similar but fall in different classes. SVM finds

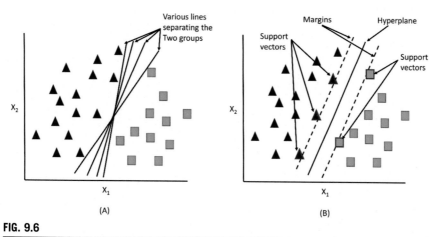

FIG. 9.6

Algorithm behind support vector machines.

such points in the space and tries to fit a line or plane, which has a maximum distance from those datapoints. These points are called support vectors, as they help in finding the decision boundary, and the algorithm is called a support vector machine.

SVM optimizes the margin of the classifier by using these support vectors, and any change in these support vectors would change the hyperplane location. An example dataset is shown in Fig. 9.6. Here, the margins are dividing the points neatly, and this margin is called hard margin. With real-world data, we use a soft margin, which allows the SVM to misclassify a few points, while, maximizing the margin and minimizing the overall error. It is up to the user to decide the degree of tolerance or the softness of the margin. A sklearn's support vector machine class is controlled using a parameter called "C". For the lower value of "C" the model will have less tolerance for misclassification, and for the high value of "C" it will have more tolerance for misclassification.

9.2.6.1 Kernel trick

We have discussed that SVMs can only classify linearly separable data, but datasets in the real-world may not always be linearly separable. Although up to some extent, using a soft margin can help us to classify data linearly, even then, the data must have a linear discriminatory boundary. For datasets that do not have any single linear boundary, the kernel trick is applied, which aims to map the low dimensional data to a higher dimension where it is linearly separable. Let us understand it using a non linearly separable one dimensional data.

Fig. 9.7A is a one-dimensional dataset that has two classes, green and blue. There is no way a point can differentiate the two classes. One dimensional dataset is a line only, and the hyperplane will be a point here, as a point has zero dimension. The goal here is to add an extra feature to the data so that we can differentiate the two classes using a line in two-dimensional space. Here the new feature is the square of the data

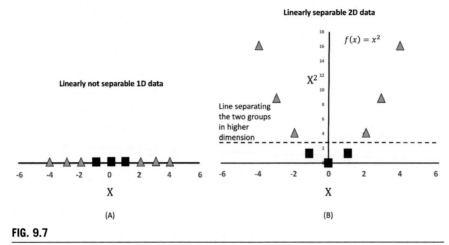

FIG. 9.7

Mapping a non linearly separable one-dimensional data to high dimension to classify it linearly.

points. Therefore, in Fig. 9.7B, we can see that introduction of this new feature made the data points linearly separable into the two different classes. Therefore, the assumption here is that, if an N-dimensional data is not linearly separable, then there is a dimension greater than N where the datapoints are linearly separable. Kernel functions are the general functions which generate new features to map these data point to higher dimensions.

Radial basis function (rbf) is one of the common kernel functions used to perform kernel trick. "rbf" is used to create new higher dimensional features by calculating the distance among all other data points to a certain point. The mathematical representation of radial basis function is

$$k\ (x_i, x_j) = \exp\ (-\gamma\| \ x_i - x_j \ \|\hat{} 2)$$

where x_i and x_j are two instances and γ (gamma) controls the effect of new features on the decision boundary. Just as "C", the regionalization parameter, gamma also needs to be tuned for optimal performance of the SVMs.

9.2.7 Neural networks

We will now address another commonly used supervised learning algorithm, Artificial Neural Networks, or ANNs, or Neural Nets, in this section. They were designed to mimic the neural networks or neurons that comprise the human brain, as the name implies. They are primarily used for statistical analysis and modeling of gathered data; their function is viewed as a complement to conventional nonlinear regression analysis models (Lecun et al., 2015). They are frequently used to solve problems that can be expressed as regression or classifications. With more than 6 decades of research behind them, neural networks have found applications in a wide variety of fields, including speech and image recognition and classification, text recognition, medical diagnosis, and fraud detection.

9.2.8 **Neural networks architecture**

The basic structure of neural networks is shown in Fig. 9.8. It is separated into three different layers: input, hidden, and output. The input layer is the first layer and it contains the input features or characteristics. The middle layer is the hidden layer, the word 'hidden' referring to the mathematical calculation processes that are not accessible and are often referred to as the black box. Diverse networks are classified according to the number of hidden layers they comprise. The ANN with many hidden layers is referred to as a deep neural network. The final layer is the output layer, which contains the network's output data.

With a large number of weights and activation functions, neural networks can accommodate both linear and nonlinear datasets. These characteristics of neural networks also make them prone to overfitting the training data. As a consequence, they require rigorous parameter tuning to achieve the best results. We will discuss more about overfitting toward the end of this chapter.

9.2.9 **Convolutional neural network**

To train an artificial neural network on image data, we must extract all the pixels from the image and pass each pixel as a variable to the neural network. Although individual pixels of an image provide information about it, images become more

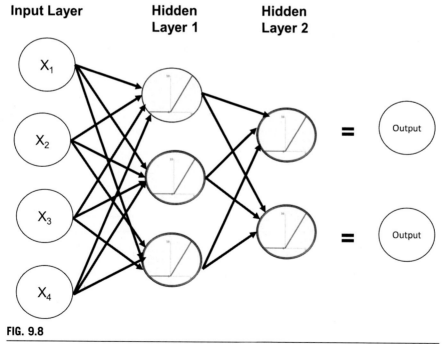

FIG. 9.8

Basic structure of neural networks.

recognizable when local pixels are combined. Convolutional neural networks (CNNs) are well-known models of deep learning for image or spatial data. CNN can be used to learn local spatial features, while ANN can be used to learn individual features. CNN learns through the use of a small window or filter. The picture is segmented into small sections, and several filters are used to learn various local features such as edges and textures. Thus, these individual image pixel features are combined to form higher-order features such as the boundaries and depths etc.

These filters are composed of convolutional matrices of numbers that can be trained or modified. With properly trained filters, CNN detects and extracts the appropriate features from image data (Fig. 9.9). Filters are trained using the back-propagation method with each iteration. By definition, convolution is a mathematical operation on two objects that results in the transformation of one object when it interacts with the other. This is often referred to as feature mapping.

Furthermore, there are two additional layers besides convolution: pooling and flattening. The pooling method moves a window across an image's pixels, fusing local features into a single feature. Pooling can take a variety of forms:

1. Max pooling: when the pixel with the highest intensity is chosen over all other pixels in the window.
2. Average pooling: where a window returns the average of all the pixels.

Pooling is primarily used to reduce the size of data. At this stage, the model retains significant information about the local pixels, such as the highest intensity or the average of all pixels, and discards other irrelevant data.

Following extraction of features from images via CNNs and pooling, these features should be fed to ANNs for classification. T he flattened layer converts these 2D features to 1D features, which are then connected to ANNs for further processing. The basic architecture of a CNN based on artificial neural networks is depicted in Fig. 9.10.

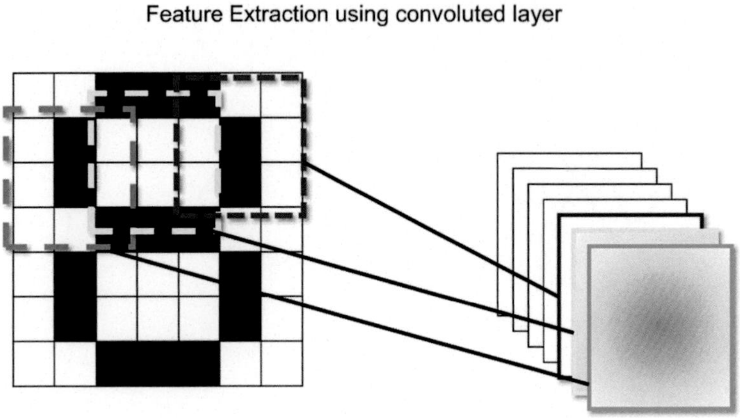

Feature Extraction using convoluted layer

FIG. 9.9

Feature extraction using CNNs.

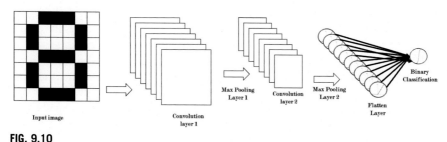

FIG. 9.10

Architecture of a CNN bases neural system.

9.2.10 **Unsupervised learning**

Unsupervised learning differs from supervised learning in that the data that we send to a computer is not labeled. To put it another way, we're not feeding the machine by hand; instead, we're making it smart enough to interpret the data we're sending it on its own. Data points with corresponding features are fed to unsupervised learning algorithms. The algorithm takes the features of the data and extracts useful bits of information based on similarities and differences within the data before trying to deduce trends.

Clustering and discovery of outliers are two extensive applications of unsupervised learning. Clustering is the most popular method of unsupervised machine learning. Clustering is the act of sorting the data into many groups. Clusters are the names of the groups segregated by machines and researchers then focus on the clusters to decipher the segregation. For example, in gene expression analysis, clustering is used to segregate the genes according to their expression patterns. Once an unsupervised algorithm group them, bioinformaticians try to understand their cumulative effects. The computer looks for data points with identical parameters in the whole data set. After that, based on each of these parameters, a data point is assigned to a special cluster.

Outlier detection, on the other hand, recognizes something that deviates from a general pattern. If you've been seeing white cars on the road for a while, seeing a red car will attract your attention. The work of the outlier detection systems is carried out in this way. This technique can be used to diagnose diseases and variations, human data entry errors, and much more.

9.2.11 **K-means clustering**

We've see examples of supervised learning in the above section so far, where labels have been used, and classifiers are build using features and labels. Unsupervised learning, on the other hand, does not have the ladled data (Altman and Krzywinski, 2017). They find hidden patterns in datasets, build clusters based on dataset similarity, and group data points that are more close to each other (called a cluster). K-Means clustering is a popular unsupervised learning algorithm, it uses the simple iterative rule to separate unlabeled datasets to form clusters. In biology, clustering has a wide range of applications, including analysis of expression results, drug repurposing, and categorization of organisms or proteins, among others.

The first step in K-means clustering is to tell the algorithms how any clusters or groups a user anticipates, i.e., the value of "K". The algorithm will cluster the data into "K" number of groups. Selecting the value of K requires knowledge of the domain or simply one can iterate the clustering process with various number of Ks and find the most suitable one based on user define parameters.

After determining the value of K, they are referred to as centroids and are randomly located within the dataset space. The sum of the squared distance between the data points and the centroid cluster (arithmetic mean of all data points belonging to that cluster) is the minimum, such that the data points closer to those randomly scattered centroids are assigned to the cluster.

In the third step, the centroids are moved to the cluster center after assigning points closer to the k clusters, as seen in Fig. 9.11. The mean of all data points in the cluster is the cluster nucleus. Closer data points are assigned to k clusters after the centroids are moved to the cluster mean, and the steps are repeated until the centroids stop moving or the movement is very slow.

Finally, we acquire well-defined k clusters from which we observe similar data points allocated to clusters to discover secret pathways and derive valuable knowledge from the dataset. Calculation of the coordinates or properties of the data points generates the nucleus of the cluster.

9.2.12 Reinforcement learning

Reinforcement learning is the process of teaching machine learning models to make a variety of decisions. The agent learns to accomplish a goal in a static, potentially tricky environment. Machine learning algorithms face a game-like scenario in reinforcement learning to solve the problem, the algorithms employ trial and error methods. Getting the ML algorithm trained is either rewarded or punished, which means it gains or loses points for the hits it makes. Its goal is to maximize the total reward or scores (Collins and Cockburn, 2020).

The idea that the user creates a layout of the game or the environment—that is, the game rules—provides the model with no tips or suggestions on how to solve the game. Starting with fully randomized trials and progressing to more sophisticated methods, it is up to the model to determine the steps to optimize the rewards. Reinforcement learning, which employs rewards and punishments through multiple testing, is possibly the most effective method for teaching an algorithm creativity. Unlike humans, artificial intelligence can derive information from thousands of concurrent gameplays if a reinforcement learning algorithm is allowed to run on a robust computing infrastructure.

Reinforcement Learning is a subset of machine learning that teaches an agent how to choose an action from an action space in each context to maximize rewards over time. Reinforcement Learning has four components:

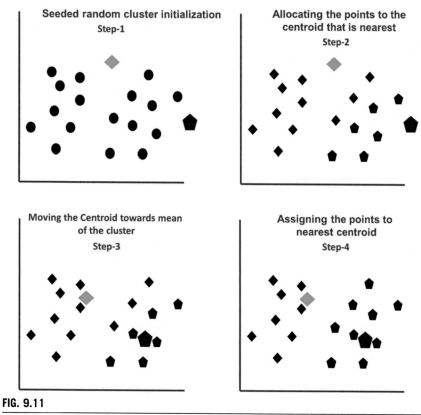

FIG. 9.11

Steps of K means clustering.

1. Agent: The ML model
2. Environment: The physical or virtual environment in which the agent operates.
3. Alternatives or steps: The agent's action that causes a change in the state of the environment.
4. Choice-based reward system: The evaluation of a particular action, which can be positive or negative.

The difference between supervised, unsupervised, and reinforcement learning is their objective and how they are achieving it. Both supervised and unsupervised learning aims to discover and learn patterns that produce relatively static results. On the other hand, RL's goal is to have a plan that tells the agent what action to take at each stage, which complicates the operation. Reinforcement Learning does not explicitly provide the correct answer; instead, the agent must learn the answer through trial and error. The agent's only metric is the reward received after the procedure is completed, indicating whether it is progressing.

In contrast, during training in supervised learning, the algorithms are given correct answers. A Reinforcement Learning agent must strike the right balance between discovering the correct steps by searching for new sources of rewards and using previously identified reward sources. Unsupervised and supervised learning systems, on the other hand, derive the solution directly from training data without considering alternative solutions. Reinforcement Learning is a multi-decision mechanism that generates a decision-making sequence based on the amount of time required to complete a task. In contrast, supervised learning is a one-decision process: one event, one prediction.

9.3 Evaluation of machine learning models

Once models have been trained, it is critical to assess their outputs or see how well they can perform (Antoniou and Mamdani, 2021). A test set is used to evaluate the performance of the models better. Typically, the entire data set is divided into two parts: 80% is used to train the machine and 20% is used to test how well the model predicts. The model is not shown the test data during training. After the algorithm has been trained, the previously unseen data is used to assess how well the algorithm performs in new data. The algorithm predicts the test data outputs, which are then compared to the original labels. The predicted and actual values are compared in a confusion matrix to measure how well the model performed (Fig. 9.12).

Various mathematical metrics can evaluate a model using the confusion matrix values, such as accuracy, precision, and recall.

9.3.1 Accuracy

The number of correctly predicted values divided by the total number of instances in the test set is the rate of accurate prediction.

Confusion Matrix

	Actually Positive (1)	Actually Negative (0)
Predicted Positive (1)	True Positives (TPs)	False Positives (FPs)
Predicted Negative (0)	False Negatives (FNs)	True Negatives (TNs)

FIG. 9.12

Confusion matrix.

$$TP + TN/TP + FP + TN + FN = \text{accuracy}$$

The ratio of true positive to the total number of instances considered positive by the model is known as precision.

The true positive rate, also known as the sensitivity of the formula, is called recollection. i.e., the total number of positive instances in the dataset of the survey divided by the number of true positive instances.

$$TP/TP + FN = \text{ranking of R F1}$$

Precision and recall are often specified a single numerical value known as the F1 score. The harmonic mean of precision and recall.

$$2 * ((\text{Precision} * \text{Recall})/(\text{Precision} + \text{Recall}) \text{ F1 Score F1}$$

9.3.2 Receiver Operating Characteristic (ROC) Curvature

The Receiver Operating Characteristic (ROC) Curve is another method for testing classifiers. It is a graph showing the difference between the true positive rate (TPR), also known as sensitivity, and the false positive rate (FPR) (1 - Specificity).

A preferred curve shows the increase in the true positive rate vs. the false positive rate. The more area there this under the curve, the more accurate the model is (Fig. 9.13).

The parameters for evaluating the regression model are very different. Since regression operates with a continuous data set, it requires sophisticated metrics to test it. Variance, mean squared error, and R squared coefficient are typical parameters for assessing regression models.

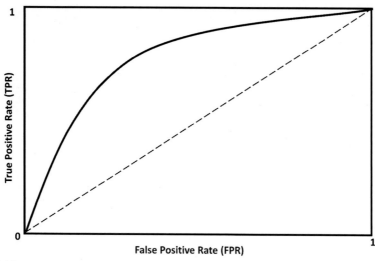

FIG. 9.13

Curve of the ROC.

9.3.3 Cross-validation

Cross-validation is a term used to describe the method of comparing two or more subjects.

As discuss above, when performing a machine learning project, we divide our dataset into two sets at random, training and test sets. This approach has a drawback: if we lose about 20% of our data for testing our model, it may not be able to train properly for real-world challenges. Some vital information will be skipped as we split the data set. When the data collection is small, the problem gets worse. To solve this problem, we use a technique called cross-validation. We divide the dataset into subsets and then we train and evaluate our model using various combinations of these subsets (Fig. 9.14). Two of these methods are K-fold validation and Leave One Out Cross Validation (LOOCV).

9.3.4 Testing and validating

When a large amount of data is available, it can be divided into three parts: training, validation, and testing sets. While training makes up the majority of the part form rests and is useful for teaching the model. The validation set is kept separate from the training data and is used for evaluating the trained model for fine-tuning its parameters, then train and evaluate again. This cycle is repeated until a model with consistent accuracy is discovered through permutations and combinations of various model parameters. The test set is used for the final evaluation because it is still hidden from the model and provides an unbiased evaluation result. Testing and

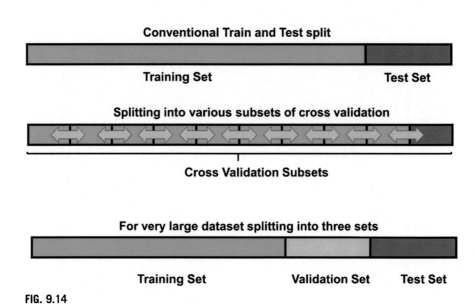

FIG. 9.14

Multiple configurations for training, validating, and testing of models.

validation on the same set of data is not recommended because, with each cycle, the model becomes biased toward the validation set due to favored parameters.

9.4 **Optimization of models**

Model optimization is a key step in improving the accuracy of the predicted outcomes. Based on the complexity of an ML algorithm, we can have various hyper parameters as we have discussed in the above sections. The purpose of fine-tuning a machine learning algorithm is to choose the best model and parameter values. In general, models are iteratively configured and validated using a number of parameter combinations. Continuous training and validation ensure that a model is optimized and performs well. Here we have discussed a few methods of optimization of ML models.

9.4.1 **Parameter searching**

There are two widely used methods for optimizing models by searching parameter:

Grid search is a method for finding the best hyperparameter combination for a model by scanning the space of all hyperparameter combinations. Depending on the type of model used, certain hyperparameters are required. Grid search is a technique for determining the optimal hyperparameters for any machine learning model. It should be noted that grid searching is computationally intensive and can take a long time to optimize a model. It iterates through all possible hyperparameter combinations and saves a model for each one. This method can be made more effective with domain knowledge by providing a limited set of parameters to evaluate.

Bergstra and Bengio (2012) proposed the concept of a system's hyperparameters being randomly searched. This is a completely different type of search than the grid approach. Rather than combing through all possible hyperparameter combinations, a randomized search selects a few sample points from the distribution and performs the calculations on those points.

9.4.2 **Ensemble methods**

Two heads are better than one means that when two people work together to solve a problem, they are more likely to succeed than when one person works alone. Ensemble methods are a class of techniques that include the development of several models and their subsequent combination to achieve improved results. Generally, ensemble models yield more precise solutions than a single model does. Voting and averaging are two of the most straightforward ensemble techniques. They are both simple to comprehend and simple to execute. Classification is accomplished by voting, while regression is performed by averaging. Both methods begin by developing multiple classification/regression models on a training dataset. Each base model can be constructed using different splits of the same training dataset and the same algorithm, or by combining the same dataset and different algorithms, or by any other approach.

9.5 Main challenges of machine learning

Machine learning enables us to make more informed, data-driven decisions more quickly than traditional methods allow. However, machine learning, like any other system, has its own set of challenges. While machine learning applications in biology are advancing rapidly, technology as a whole has a long way to go. When it comes to developing a machine learning solution, there are numerous obstacles and issues to overcome. With this in mind, let us examine some of the challenges faced by applied machine learning projects.

9.5.1 Insufficient quantity of training data

Having sufficient data to train the algorithm is critical to the machine learning project's success. Many machine learning programs may fail to approximate the real-world scenario due to a lack of data. As a result, data collection is the most critical component of any machine learning project. The amount of data depends on the complexity of the problem and the chosen ML algorithm. While high throughput technologies have significantly increased the amount of data available for biological studies, the integration of divergent types of data remains a limitation for big data in biology. High-throughput data from transcriptomics, proteomics, metabolomics, and genomics must be integrated into a meaningful large dataset. It remains a challenge that, despite the production of astronomical amounts of data in the biological domain, the overall mechanism of a system such as phenotypes remains difficult to comprehend due to a lack of data integration.

9.5.2 Non-representative training data

Training data must be a representative sample of the population. A sample data is a subset of the population's data points used to train a model. The sample must accurately reflect the population. To generalize the output of a machine learning model, it must be trained on well-represented data; otherwise, it will encounter a blind spot for predictions on data with few representations.

Sampling noise, also known as non-representative data, occurs when the sample size is insufficient, but even large samples can be non-representative if the sampling procedure is inaccurate. This is frequently referred to as sampling biased. It is critical to note that when sampling bias is reduced, the variance increases, preventing the model from generalizing to unseen data; when variance is reduced, the bias increases. This phenomenon also referred to as the bias-variance trade-off; one should balance these two parameters.

9.5.3 Quality of data

Machine learning algorithms pose a huge challenge when it comes to data consistency. These systems require high-quality data to prevent over or under estimation during training and testing cycles. As a result, the consistency of the data used in any machine learning system has a huge impact on its development. Consequently, even small variations in the training data can result in major alteration in the algorithm's output. Complex and dynamic problems require not only large quantities of data but also diverse and informative data. While curating data sets, the missing, misinterpreted, and inaccurate values may affect data quality.

9.5.4 Irrelevant features

The independent variables in any machine learning problem are the features. The output, which is a dependent variable, is predicted using these independent variables. During training, the majority of machine learning algorithms derive an approximate function from these features. As a result, it is critical to select relevant features for inclusion in a machine learning algorithm. Relevant features can be used to direct the algorithm toward a more desirable output. Inappropriate features mislead an algorithm, increase the size of the data, and complicate the problem unnecessarily. In many cases, feature redundancy can also result in an increase in data size. Two or more highly correlated features have little "value" in terms of training because the presence of one can always (or almost always) be used to determine the presence of the other. If this is the case, there is no reason to include both features, as their combined effect on the predictions will be negligible.

Sometimes the simplest solution is the best. To be more specific, we can use dimensionality reduction while retaining variance, as in PCA, by using the principal components as features. Furthermore, because more features introduce more noise, we can use a variety of methodologies and techniques to select a subset of the feature space to aid our models in performing better.

We have discussed the difficulties and limitations of machine learning as a result of the scarcity of useful data thus far. Data has become the new gold as a result of advancements in machine learning. But there are some limitations in algorithms themselves, which we will discuss in the next section.

9.5.5 Overfitting or underfitting on training data

Overfitting occurs when a model performs exceptionally well during training but performs poorly on test data. It takes place when an algorithm acquires an excessive amount of knowledge about the training data. When a model acquires unnecessary extra details and noise from training data, it has a negative impact on the model's performance on test data. As a result, the model will be incapable of generalizing the real-world situation. Generally, nonlinear machine learning algorithms with a large number of parameters overfit the training data. Overfitting can be avoided by fine-tuning the parameters.

When a system cannot precisely capture the relationships between features and output variables, this is referred to as underfitting. This impairs the algorithm's ability to decode the data's underlying pattern. Underfitting is a good indication that the algorithm is unsuitable for the given dataset. Underfitting is typical in situations where there is insufficient data to train the algorithm. Fewer data points result in an inconsistency in the model's training, which then fails to perform well on the test dataset. Increasing the amount of data available to the algorithm may be one way to ensure that it has enough information to identify general trends and patterns. Another way to tackle underfitting is to look for other algorithms.

References

Altman, N., Krzywinski, M., 2017. Points of significance: Clustering. Nat. Methods 14, 545–546. https://doi.org/10.1038/NMETH.4299.

Altman, N., Krzywinski, M., 2015. Points of significance: Simple linear regression. Nat. Methods 12, 999–1000. https://doi.org/10.1038/NMETH.3627.

Antoniou, T., Mamdani, M., 2021. Evaluation of machine learning solutions in medicine. CMAJ (Can. Med. Assoc. J.) 193, E1425–E1429. https://doi.org/10.1503/CMAJ.210036/TAB-RELATED-CONTENT.

Bergstra, J., Bengio, Y., 2012. Random search for hyper-parameter optimization. JMLR 13 (10), 281–305.

Collins, A.G.E., Cockburn, J., 2020. Beyond dichotomies in reinforcement learning. Nat. Rev. Neurosci. 21 (10), 576–586. https://doi.org/10.1038/s41583-020-0355-6.

Greener, J.G., Kandathil, S.M., Moffat, L., Jones, D.T., 2021. A guide to machine learning for biologists. Nat. Rev. Mol. Cell Biol. 23 (1), 40–55. https://doi.org/10.1038/s41580-021-00407-0.

Kotsiantis, S.B., 2011. Decision trees: A recent overview. Artif. Intell. Rev. 39 (4), 261–283. https://doi.org/10.1007/S10462-011-9272-4.

Larrañaga, P., Calvo, B., Santana, R., Bielza, C., Galdiano, J., Inza, I., Lozano, J.A., Armañanzas, R., Santafé, G., Pérez, A., Robles, V., 2006. Machine learning in bioinformatics. Briefings Bioinf. 7, 86–112. https://doi.org/10.1093/BIB/BBK007.

Lecun, Y., Bengio, Y., Hinton, G., 2015. Deep learning. Nature 521 (7553), 436–444. https://doi.org/10.1038/nature14539.

Lever, J., Krzywinski, M., Altman, N., 2016. Points of significance: Logistic regression. Nat. Methods 13, 541–542. https://doi.org/10.1038/NMETH.3904.

Parry, R.M., Jones, W., Stokes, T.H., Phan, J.H., Moffitt, R.A., Fang, H., Shi, L., Oberthuer, A., Fischer, M., Tong, W., Wang, M.D., 2010. k-Nearest neighbor models for microarray gene expression analysis and clinical outcome prediction. Pharmacogenom. J. 10 (4), 292–309. https://doi.org/10.1038/tpj.2010.56.

Wang, P.W., Lin, C.J., 2014. Support vector machines. In: Data Classification: Algorithms and Applications. https://doi.org/10.1201/b17320.

Systems and network biology

10

10.1 Introduction

Biological systems exhibit a high degree of complexity, which is achieved through multiple layers of hierarchy. At the heart of this assembly is the genome, which contains information necessary for the creation of molecules and the execution of a variety of processes. Following the completion of the human genome project, the genomic aspect of biology has advanced. To make use of the massive amounts of biological data generated by high-throughput techniques, novel methods for hypothesis-driven research are being developed. Now, computational methods for large-scale data analysis, as well as modeling and simulations of complex sequences, are being used. One of the most significant outcomes of the Human Genome Project was the acceleration of scientists' adoption of a new systems approach. Systems biology examines the properties and relationships between the components of a biological system during its process. This data can be combined, visualized as graphs, and modeled computationally. The Human Genome Project pioneered a new method of biology called discovery science. This defines the genetic components of humans and other organisms, views biology as a branch of information science, and provides high-throughput techniques for system understanding and the development of newer computational techniques. For more than a century, scientists have studied the individual components of cells and their functions. Despite its enormous success, it was discovered that a single molecule is rarely capable of performing a distinct biological function. The majority of the characteristics of biological systems have been discovered to be the result of complex interactions between various cellular components, including RNA, DNA, proteins, and other molecules. Thus, a significant challenge for biologists in the 21st century is to comprehend the structural and dynamic properties of this intricate web of interactions. The development of new high-throughput techniques, such as microarrays, enables us to monitor the status of cellular components at any time. Protein chips or yeast two-hybrid screening can reveal how and when these molecules interact (Altaf-Ul-Amin et al., 2014).

All About Bioinformatics. https://doi.org/10.1016/B978-0-443-15250-4.00011-3

10.2 Network theory

The study of network theory is concerned with the representation of asymmetric or symmetric relationships between objects in a graph. In mathematics and computer science, network theory is a subfield of graph theory. Network theory is applied in a variety of fields, including physics, biology, computer science, and finance. This theory is applicable to social networks, the internet, and metabolic networks, among others. As a result of molecular interactions, various types of networks (e.g., transcription-regulatory, signaling networks) emerge. These networks cannot exist in isolation. These networks combine to form a "network of networks," which determines the cell's behavior. A significant challenge is to integrate theoretical and experimental data in order to map, comprehend, and model the characteristics of various networks that contribute to the behavior of these cells. The theory of complex networks has aided in our understanding of the mechanisms governing the formation and evolution of various social and technological networks. This study established that the architecture of cellular networks is comparable to that of complex systems such as society, the internet, and computer chips. This similarity demonstrated that the same laws govern the majority of complex networks found in nature, ranging from large non-biological systems to cellular systems (Barabási and Oltvai, 2004; O'Connor, 1992).

10.3 Graph theory

Leonard Euler, a mathematician, introduced graph theory for the first time. It was used to assist Königsberg in resolving a problem. Seven bridges connected four islands in this problem. Nobody could find a path that crossed all four islands and each bridge at the same time. As a result, people believed that no such path existed, but this could not be mathematically established. Euler found the solution to this problem by focusing exclusively on the relationship between the land masses and disregarding the distances and shapes of the paths. Euler demonstrated that such a path was not possible using topological features and graphs. Euler's concept serves as the foundation for graph theory (Fig. 10.1).

In mathematics, graph theory is the study of graphs that are used to generate a model of the pairwise relationship between objects. These graphs are made up of vertices called nodes that are connected by edges called links.

Networks can be used to represent a variety of different types of data. These networks can contain a variety of different entities, such as genes or proteins. Edges represent the data associated with the connection between these nodes.

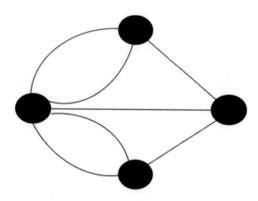

FIG. 10.1

Königsberg's problem of four islands connected by bridges.

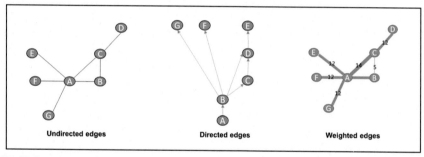

FIG. 10.2

Types of network edges.

10.4 **Features of biological networks**

10.4.1 **The various types of network edges**

Numerous types of analysis can be performed using the information provided by the edges.

There are three primary types of edges:

(1) Undirected Edges: These edges represent relationships between nodes but do not indicate the direction of flow. The relationship merely demonstrates that A binds B via the evidence. These types of edges are frequently found in networks of protein—protein interactions.

(2) **Directed Edges:** These edges are useful for hierarchically organizing a network and depict the direction of signal flow. These types of edges are frequently found in networks regulating gene expression or metabolism.

(3) **Weighted Edges:** Weights can be applied to both undirected and directed edges. These weights can be used to display a variety of values, including sequence similarity between genes and interaction reliability. Additionally, other topological features can lend weight to the edges (Fig. 10.2)

10.4.2 Network measures

Network biology generates quantifiable networks for the purpose of characterizing various biological systems. Some fundamental network measures can be used to compare and characterize various networks.

Degree (k): A node's most fundamental property is its degree (also known as connectivity). The degree of a node indicates the number of connections it has with other nodes. For instance, Fig. 10.2 illustrates an undirected network with node A having $k = 5$. In some networks, each link has a fixed direction. In these networks, k_{in} (incoming degree) represents the number of links leading to a particular node, while k_{out} (outgoing degree) represents the number of links leading away from a node. For instance, in Fig. 10.2, Node B has a k_{out} of 4 and a k_{in} of 1. For undirected networks with L links and N nodes, the average degree is equal to $2L/N$.

Degree distribution [P(k)]: This attribute indicates the probability that a given node will have k connections. The degree distribution is calculated by dividing the total number of nodes N by the number of nodes N(k). P allows for the differentiation of various classes of networks (k).

Scale-free networks: The majority of networks are scale-free, which means that their degree distribution is roughly equal to that of a power law with exponent. Hubs with lower values play a more critical role in the network. Hubs with fewer than three nodes are not considered relevant. Between 2 and 3 hubs, a hierarchy of hubs can be created. The majority of nodes in these networks are connected to a small number of neighbors. These networks' hubs contribute to their high connectivity. The hubs in Fig. 10.3A are highlighted in orange.

Path length is defined as the number of connections between two nodes. In directed networks, the smallest number of links between nodes B and A differs from the smallest number of links between nodes A and B. For instance, in Fig. 10.3B, AB equals three and BA equals one. At times, no direct path between two nodes may exist. For instance, In Fig. 10.3B, there is no path from A to C, but there is one from C to A. The average of the path lengths between all the nodes in a network is referred to as the mean path length, or l>. This metric indicates the network's overall navigability.

Coefficient of clustering or transitivity: The clustering coefficient quantifies clusters (closely linked nodes in a network). In a cluster, nodes are more connected to one another than in the rest of the network. Fig. 10.3C, illustrates the various clusters. If node X is connected to node Y and node Y is connected to node Z, there is a

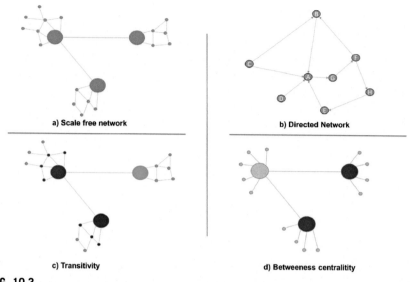

a) Scale free network

b) Directed Network

c) Transitivity

d) Betweeness centralitity

FIG. 10.3

Some network measurements.

high probability that node X is connected to node Z. This observation can be quantified using the equation $CA = 2nAk(k-1)$, where nA denotes the number of connections between kA neighbors of node A. As a result, CA indicates the total number of triangles that pass-through node A. The average clustering coefficient, or C>, can be used to calculate the tendency for cluster formation within a network. $C(k)$ returns the mean clustering coefficient for each node connected by k.

Centralities: A network's centrality indicates the importance of an edge or node in terms of information flow or connectivity. Centrality is affected by a variety of factors, including a node's degree. Fig. 10.3D, illustrates the centrality of betweenness. Central nodes are colored differently, and the degree of nodes is determined by their node size (Table 10.1).

Table 10.1 The different measures of a network and their definitions.

Network measure	Definition
Degree	It indicates the number of connections from a nodes.
Degree distribution	The likelihood of a particular node having k connections.
Scale-free networks	In these networks degree distribution is approximately equal to a power law. $P(k) \sim k^{-\gamma}$
Path length	The number of links present between two nodes.
Clustering coefficient	Clustering coefficient gives a measure of clusters in a network.
Centralities	Centrality of a network predicts the importance of an edge or a node to the flow of information or to the connectivity.

10.4.3 Network models

Network models are critical for comprehending complex networks and explaining their properties. The three most critical biological networks are as follows:

Random Networks: The ER model represents a random network that begins with N nodes and assigns each pair of nodes a probability p. This generates a graph with links distributed randomly. The node degrees are then distributed according to a Poisson distribution, indicating that the majority of nodes are connected by the same number of nodes. P(k) decreases incrementally in regions with high k values, indicating that the majority of nodes are present near the average value. Because C(k) is not dependent on the degree of a node, it is represented by a horizontal line as k's function. The mean path length of a network increases in logarithmic proportion to the network's size.

Scale-free Network models: These are defined by a power-law degree distribution; the probability of a node having k links can be calculated as follows:

$$P(k) \sim k^{-\gamma}$$

In this case, the degree exponent equals. When compared to random graphs, highly connected nodes have a higher probability of being present in these networks. The properties of these networks are determined by hubs (small numbers of nodes with a large number of connections). According to Barabási– Albert's model of a scale-free network, the network has a power-law degree distribution defined by the degree exponent = 3. Such models lack inherent modularity; the C(k) here is not dependent on k. The degree exponents of these networks, which are observed in networks, are smaller than the degree exponents of random networks.

Hierarchical network models: Hierarchical network models are iterative approaches for generating networks that may recreate the unique qualities of a scale-free topology while still exhibiting a high degree of node clustering. These qualities are abundant in nature, ranging from biology to language to certain social networks. To justify the presence of modules, it is assumed that clusters integrate in a repetitive manner, resulting in the formation of hierarchical networks. These networks are constructed using small densely connected clusters of nodes. This is followed by the generation of duplicate modules, with the external node of each cluster being connected to the central node of the previous cluster, resulting in the formation of a large node module. Identical modules are constructed, and then the peripheral nodes connected to the central node module. Nodes with low connectivity are concentrated in dense regions of the network in a hierarchical structure. These densely populated areas are linked by a few hubs (Fig. 10.4).

10.5 Types of biological networks

In a cell model, various forms are utilized to represent various types of information. The information gathered from nodes and edges varies according to the type of data utilized to construct the network. Different data types will result in different network

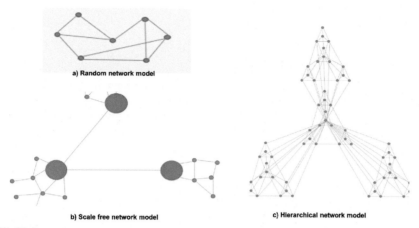

FIG. 10.4

Types of network models.

characteristics, including structure and connectivity. Multiple pieces of information can be conveyed in these networks via edges and nodes (Fig. 10.5).

The major classifications of biological networks are as follows:

(1) Cell signaling networks
(2) Gene/transcription regulation networks
(3) Genetic interaction networks

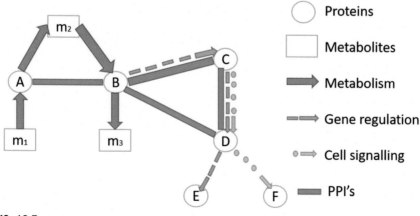

FIG. 10.5

Representation of the different biological processes as a network.

(4) Metabolic networks
(5) Protein—protein interaction networks

10.5.1 Cell signaling networks

This network is a complex model of communication that regulates a variety of cell actions and activities. Immunity, repair, and development are all included. The cell in this network converts one type of signal to another. Serious diseases (e.g., diabetes, cancer) can be triggered by cellular signaling abnormalities. A set of metabolic events occurring within a cell and the alterations are occur as a result of receptor activation via signaling pathways. Proteins act as nodes in these networks, which are connected by directed edges.

Significant signaling pathways include the following:

MAPK pathway (mitogen-activated protein kinase): MAPK is a protein kinase enzyme that is capable of phosphorylating a variety of proteins, including transcription factors. For instance, MYC, an oncogene transcription factor, is expressed in a broad range of human malignancies. MAPK can phosphorylate MYC, hence altering the cell cycle and gene transcription. The MAPK pathway can be activated by the EGFR (Epidermal growth factor receptor). Cancer can be caused by abnormalities in the expression.

Signaling pathway of the hedgehog: This route is a critical regulator of development in animals. From flies to humans, this pathway is retained. This pathway enables the establishment of the flybody plan. During metamorphosis and development, this route is critical.

Pathway of TGF-beta signaling: This pathway is involved in cell proliferation, differentiation, and apoptosis.

10.5.2 Gene/transcription regulation networks

These networks serve as a model for the regulation of gene expression. Gene regulation is the process by which instructions contained in genes are transformed to products (such as RNA or protein). Gene regulation is responsible for the structure and function of cells. Cell differentiation, for example, can transform a cell into a specialized form. Through morphogenesis, an organism can acquire a shape.

Gene expression and repression are presented identically in these networks. Model organisms are examined in order to gain a better understanding of how other creatures function. D. melanogaster, *E. coli*, and *S. cerevisiae* are among these model organisms.

10.5.3 Genetic interaction networks

Genetic interactions are a cooperative phenomena in which mutations in two or more genes result in a phenotype that is different than expected when the effects of each mutation are taken together. These networks do not depict a physical connection between the genes, but rather a functional connection. Genes are represented as nodes, and their interconnections are represented by their edges. It is possible to deduce the direction of the edges.

10.5.4 Metabolic networks

These are the biological reactions that take place within a cell. These networks are reconstructed using data from experiments and the genomic sequence. There are networks accessible for a wide variety of creatures, from bacteria to people. These networks can be used to simulate and analyze metabolism. Metabolism is a collection of biological events that assist organisms in maintaining their structure, responding to environmental changes, developing, and multiplying. Metabolic pathways are a set of chemical events occurring within the cell that are catalyzed by enzymes. This can result in the formation of a product or can initiate the beginning of further metabolic pathways. Numerous metabolic pathways coexisting in a cell might form a metabolic network. Metabolic pathways are the major responses that maintain the homeostasis of an organism. The paths between the substrates and the enzymes contain directed edges. As a result, these enzymes and substrates are represented as nodes in the network, whereas directed edges depict the reactions.

10.5.5 Protein—protein interaction networks

Protein—protein interactions are required for the majority of biological functions. As a result, understanding these PPIs is critical for understanding how cells' physiology changes during various disorders. Because drugs have an effect on these PPI networks, understanding PPIs is critical for drug development. In general, protein—protein interactions (PPIs) relate to the physical interactions or binding between proteins. These interactions are very specialized, occur in defined binding areas, and have a defined function. PPI data can be used to represent both steady and transient interactions. Protein complexes have stable interactions (e.g., hemoglobin). Modifications to proteins are temporary interactions (e.g., kinases). The interactome's dynamic component is composed of transient interactions. PPI data can be used to:

- Assign proteins functions
- Recognize minor details regarding signaling pathways
- Determine the relationships between the proteins contained in complexes such as the proteasome.

Interactome: The interactome is a term that refers to the collection of all PPIs found in a cell, an organism, or any other environment. The advancement of PPI screening tools such as the yeast two-hybrid experiment and mass spectrometry has resulted in a massive rise in the amount of PPI data and the production of more complex interactomes.

The properties of PPI networks include the following:

The small world effect in PPI networks shows that there is a high degree of interaction among the proteins. A network's diameter (the largest distance between two nodes) is always small. Between two nodes, the distance is always less than six steps. This is referred to as the "6° of separation" principle. The degree of connection is critical biologically, as it enables the rapid and efficient flow of impulses within a network. Although networks are highly coupled, alterations in a single gene or protein have little effect on the network due to the network's resiliency.

Scale-free networks: In these networks, the majority of proteins (nodes) are connected to only a few other proteins. A few proteins (hubs) are linked to a vast number of proteins (nodes) in the network. The networks can be structured in a preferential attachment approach. This model is based on the notion that when a network is built, edges are attached preferentially to nodes with the greatest degrees. This principle aids in the establishment and expansion of the network.

Due to their scale-free nature, PPI networks exhibit the following characteristics:

Stability: When a random failure occurs, the probability of harming a hub is low due to the low degree of connection of the majority of proteins. Due to the availability of multiple hubs, a network will not lose connectivity if a single hub fails.

Invariant to scale change: Regardless of the amount of nodes or edges in a network, it will always stay stable. Regardless of the network's size, hubs produce small world effects.

Vulnerability to a targeted attack: When a small number of networks in a hub fail, the network degrades into a collection of isolated graphs. The hubs contain both lethal and vital genes. For instance, some proteins associated with cancer are hub proteins (e.g., p53 protein).

Transitivity: Another critical aspect of these networks is their modularity. The clustering coefficient or transitivity of a network measures the inclination of nodes to cluster together. A higher transitivity score indicates that a collection of nodes is highly connected (communities). It is critical to locate these communities within biological networks, as they can reveal protein complexes and functional modules. The module is a term that refers to an interchangeable functional unit. These modules are network contents with known relationships to other network contents. The critical property of a module is that its internal properties remain constant regardless of the context in which it is utilized. Modules simplify biological networks by dividing them into smaller, functional components that may be perceived as a unified unit. Topological network studies can aid in locating and comprehending these modules.

A complex of proteins can be thought of as a module with stable interactions between the proteins that result in a fixed configuration at a particular time and location. These complexes are representations of multi-protein models performing prescribed functions. A broader version of these modules does not require protein binding, as their functions remain constant when used in a different setting. Additionally, knowledge of the modules aids in the comprehension of their interactions with one another and with proteins (Fig. 10.6).

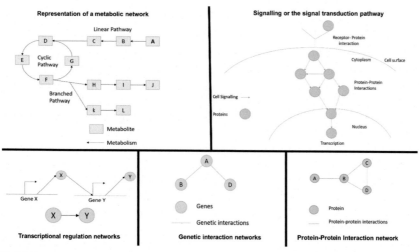

FIG. 10.6

Classifications of biological networks.

10.6 Sources of data for biological networks

The data can be obtained from different sources for generating biological networks such as:

Manual literature curation: Curators or domain experts can review previously published information and store it in databases. This can result in high-quality, well-presented data. However, curation takes a significant amount of time and money, limiting the size of databases.

High-throughput datasets: Experiments such as mass spectrometry identification and yeast two-hybrid are generating large volumes of data, such as the PPI datasets. Although large, ordered datasets are generated, the approach used introduces bias, and the quality of these datasets may vary.

Computational techniques: A variety of strategies are used to anticipate previously unknown associations between items based on current experimental evidence. For example, if orthologues exist, protein interactions in mice can be predicted by examining protein interactions in humans. This results in the creation of a tool for broadening the scope of experimental data. However, the datasets generated by these methods are noisier than the ones discussed previously.

Extracting relationships from the literature via text mining: Various machine learning techniques can be used to extract relationships from the literature. While this strategy can increase data coverage, NLP is time consuming to process and the resulting results are frequently noisy.

10.7 Gene ontology for network analysis

The initial output of genomics experiments is a list of genes that significantly alter the experimental conditions. The first step in investigating datasets is functional enrichment analysis. This determines whether the gene list is statistically enriched for particular biochemical functions and processes. The Gene Ontology (GO) consortium develops a controlled vocabulary of terms that can be used to describe genes and their products as cellular components, biological processes, and molecular functions. Utilizing the open access tools available on the GO website, an enrichment analysis could be conducted using GO. These analysis techniques examine a list of genes for the presence of GO terms that occur more frequently than expected by chance in the query list. The presence of over-represented terms indicates that unknown processes are controlled differentially and preferentially in a given condition. Both GO's strength and weakness stem from its hierarchical structure. Despite efforts to comprehend the architecture of GO enrichment analyses, determining the level of hierarchy that results in statistical enrichment is challenging. By and large, the major enriched terms encompass broader functional categories that are less useful for comprehending novel functions. Cells have pathways that are biochemical systems that aid in the conversion of signals to output. Enrichment analysis on the basis of a pathway provides information that is more pertinent and understandable in the context of critical processes occurring under a given condition. Numerous techniques for pathway analysis are available, including DAVID, InnateDB, KEGG, and Reactome; quantitative techniques rely on gene set enrichment. Current techniques attempt to take advantage of the fact that not all genes are equally capable of distinguishing between different pathways. As a result, pathway analysis can provide us with a wealth of information about previously unknown relationships between genes or pathways (Yon Rhee et al., 2008).

While pathway analysis tools are extremely powerful, they do have the following limitations:

- The majority of genes do not have a pathway annotation. For example, more than 85% of human genes in the Ensembl database are not mapped to KEGG pathways.
- There is a significant bias in favor of the more well-studied pathways.

10.8 Analysis of biological networks and interactomes

Network biology is a rapidly growing field of research that demonstrates that biochemical processes are not regulated by isolated proteins or straight pathways, but rather by a complex system of networks of interacting molecules. Understanding molecular interaction networks can aid in the development of biochemical processes and identifying the critical topological properties and nodes required for their control, which are important for explaining the complex symptoms of diseases.

According to network theory, disease phenotypes are not caused by mutations in single genes acting independently, but rather by changes in a gene's network context. Thus, understanding disease pathology and developing effective targeted therapeutics requires a thorough understanding of pathologically altered interaction networks. Network analysis is a time-efficient technique that is comparable to more traditional methods. The advantages of this technique are that it is more data intensive and less constrained by the limitations of currently used functional annotations, as a result of the existence of proteome level interactome maps for various species, such as humans. Network analyses are less biased toward well-characterized pathways and incorporate a broader variety of proteins and genes. The interactome can be visualized and comprehended by constructing a network or graph in which each entity (metabolite, miRNA, protein, or gene) is represented by a node and its interactions with other entities by edges connecting the nodes. Networks can contain a variety of nodes and edges and are frequently used to visualize and interpret a variety of different types of entities and their relationships concurrently. This can represent a highly accurate and comprehensive representation of a biological system. Attributes of nodes or edges can aid in the integration of biological data associated with the node's edges (e.g., confidence score) or (e.g., gene expression). Two distinct methods are used to conduct network analysis on a list of genes. The first method superimposes genomics data (for example, gene expression) on preestablished interaction networks (e.g., open access PPI data). The second method is to discover the network using data generated experimentally (Kovács et al., 2019).

10.9 Interaction network construction using a gene list

The type and source of data are critical considerations when constructing interaction networks. When determining the quality or type of interaction data, researchers must proceed cautiously. The IMEx (International Molecular Exchange) consortium's databases promote manual curation of experimentally generated data from the research literature. Meta-databases aggregate and repackage data from primary sources and make it accessible through a portal. A few databases incorporate computational interactions to augment the experimentally generated data. This step is critical for enriching the experimentally generated sparse interaction network. Researchers can obtain more accurate results by comparing experimentally generated data to data generated computationally. Primary interaction databases exhibit a low degree of overlap in the data they provide. This is done on purpose, as the developers of the IMEx database wish to avoid duplication in the manual curation method. This also results in the omission of supportive interaction data when a single database is used as the source. Researchers can search multiple databases concurrently using web services such as PSICQUIC (Aranda et al., 2011).

Prior to conducting network analysis, certain information must be considered. Extracted interactions can be physical (e.g., PPI), regulatory (e.g., miRNA-mRNA), or biochemical in nature (e.g., phosphorylations). While combining different types of interactions can be beneficial, we should proceed cautiously because the edges in such a network would vary significantly, and this factor should be considered when analyzing the data. For example, PPIs typically have undirected edges and can access information about protein complexes, whereas biochemical interactions are directed and can depict data flow. Another aspect of protein interactions is that they can be detected using Mass Spectrometry and Affinity Purification, but these techniques are unable to distinguish between indirect and direct interactions, despite the fact that they are depicted as interactions in open access databases. Confidence in a specific interaction varies significantly depending on the experiment used to determine it. Massive amounts of data about the interactome could be generated using high-throughput technologies, such as yeast 2-hybrids. However, this data has a high rate of false negatives and positives. Meanwhile, interactions extracted from the literature inculcate greater confidence but favor well-characterized biological processes and pathways. Numerous methods have been developed to generate confidence scores, which are displayed on networks as edge weights. Interactomes derived from databases are a static representation of all the interactions associated with a given input list. Certain interactions are context-dependent (e.g., occurring in a particular type of cell, under certain conditions, or for a particular protein isoform). However, the literature and interaction databases are deficient in high-throughput interactome data. If this analysis is restricted to specific contexts, the vast majority of data will be omitted. Researchers are now integrating multiple types of contextual data, including gene and protein expression, to identify the most probable edges and nodes in a subnetwork.

10.10 Data analysis tools

10.10.1 The InnateDB

The InnateDB database (Breuer et al., 2013) contains a large number of experimentally verified bovine, mouse, and human interactions and pathway annotations that were compiled from publicly available interaction and pathway databases. Along with the integrated data, the InnateDB curation group has annotated a large number of immune-related reactions through a review of the biomedical literature. The interactions in InnateDB have been curated according to MIMIx standards. Additionally, InnateDB offers integrated bioinformatics tools such as ontology and pathway analysis, network analysis and visualization, as well as the ability to take and interpret user-supplied gene expression data in a pathway or network context. InnateDB also contains the entire mouse genome. In the first step, a list of genes is chosen. After that, human Ensembl gene IDs are mapped to selected gene IDs. This can be accomplished by selecting the "Data Analysis" menu and then the "Network Analysis" tab. Following that, we can upload the gene list using the "Upload Data" button. Alternatively, you can upload this data in the form of a spreadsheet or text file. More than 10 different quantities can be used to quantify data associated with genes, which can be integrated as different node attributes in a network visualization. InnateDB provides a variety of filters for locating the interactions in the developed network. The default option returns all interactions that contain at least one gene from the gene list. This enables us to identify nodes in networks that have not been identified experimentally but are capable of interacting with these genes or proteins. The user can choose to view only interactions between the input genes or interactions within a particular pathway of interest. One can choose between predicted interactions and manually annotated interactions. The user is then presented with a table of reviews for the uploaded list. After that, the query data can be used to construct networks.

10.10.2 Visualization and download of networks

The networks generated by InnateDB can be viewed using a variety of different tools. The Innate DB results page is visualized using the CerebralWeb app (Frias et al., 2015), which layers nodes according to their subcellular location. A Cerebral plugin can be used to analyze interaction networks. Additionally, third-party software such as BioLayout Express 3D and CyOog plugins can be used. BioLayout Express 3D enables the visualization of large networks in 3D or 2D. The CyOog plugin analyses and reduces the complexity of networks by utilizing Power Graph. InnateDB networks can be downloaded in a variety of formats (Fig. 10.7).

10.10.3 Enrichr

Enrichr (Chen et al., 2013) is a web and mobile application that combines gene set libraries, a novel technique for ranking enriched terms, and multiple methods for visualizing results. Enricher is an HTML5-based application. Users are able to share

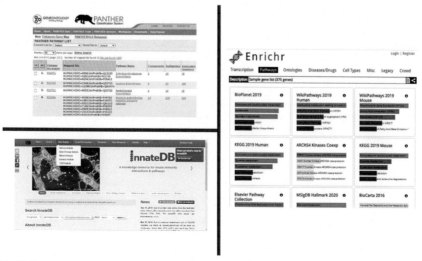

FIG. 10.7

Webpage of InnateDB, Enrichr and PANTHER.

results and view them in publication-ready formats. Enrichr is a simple web-based enrichment analysis tool that generates various visualisations for the combined functions of an input gene list (Fig. 10.7).

10.10.3.1 PANTHER

PANTHER (Thomas et al., 2022) integrates pathways, ontologies, gene functions, and statistical analysis methods to analyze large amounts of data from gene expression, proteomics, and sequencing experiments. This system utilizes 82 genomes organized into subfamilies, families, and phylogenetic trees to visualize the relationships between genes, MSA, and HMM models. Numerous classification schemes are used to classify genes, including annotated subfamilies and families, as well as sequences associated with pathways. The PANTHER website contains a variety of tools that enable users to search for and screen gene functions, as well as analyze experimental data using a variety of statistical methods (Fig. 10.7).

10.10.3.2 GESA

Although RNA expression analysis has become a standard technique in biomedical research, extracting useful biological information remains a significant challenge. GSEA is a technique for analyzing and comprehending expression data. This technique focuses on a group of genes that share a common location, function, or regulatory role. GSEA can detect similarities in clinical data collected from patients across studies, revealing shared pathways. The GSEA technique is an open-source package that includes a database containing several standard gene sets (Fig. 10.8).

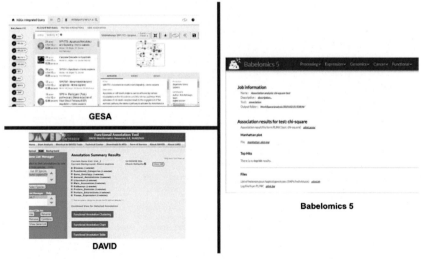

GESA

DAVID

Babelomics 5

FIG. 10.8

Webpage of GESA, DAVID and Babelomics 5.

10.10.3.3 DAVID

The DAVID (Huang et al., 2007) bioinformatics resources include a knowledgebase and tools for mining biological information from large protein or gene lists. DAVID, a data mining environment, is capable of analyzing gene lists generated during genomic experiments. This process begins with the upload of a list of genes with multiple gene identifiers that are analyzed using pathway or text mining tools such as gene functional classification, functional charts, or tables. This method gives researchers with biological information contained in gene lists enriched by genome studies (Fig. 10.8).

10.10.3.4 Babelomics 5

The Babelomics 5 (Medina et al., 2010) tool accepts a list of genes or proteins and maps them to a reference interactome. The interactome can be constructed using any user-defined interactome, including human interactomes. Following the list's mapping, Babelomics 5 determines the values of several other parameters using the least connected networks and interactomes defined by proteins. Comparing separate protein lists allows for the detection of significant changes in their parameter distributions (Fig. 10.8).

10.11 Network visualization tools

10.11.1 Cytoscape

Cytoscape (Shannon et al., 2003) is an open access environment for combining, visualizing, and querying networks. Cytoscape's main software part gives the ability to input and output data, integrates interactions, network, data visualization, filtering and querying tools. Cytoscape's VizMapper helps give visual mapping to attributes, controlling visualization of edges and nodes on the basis of their molecular states. These mappings help to overlay many data types in networks. Cytoscape was created in Java and distributed as a free software. It has been integrated with other applications (e.g., geWorkbench) and tools, websites (e.g., network image generator), databases (e.g., BIND, MiMI). Commercial companies like GeneGO, Genespring, Agilent have utilized this software. The core of Cytoscape can be expanded through plugin structures, allowing fast modeling, development of analysis and features. Many third-party programmers are involved in developing plugins for Cytoscape, due to the popularity of Cytoscape as a free environment. 74 open access plugins have been produced since 2004, 46 of these are compatible with the latest versions of Cytoscape (Fig. 10.9).

10.11.2 NAViGaTOR

NAViGaTOR (Brown et al., 2009) is a free package for network visualization. It can be used as an alternative to cytoscape. It uses openGL to accelerate graphics and allows quick rendering and visualization of huge networks. Options to visualize

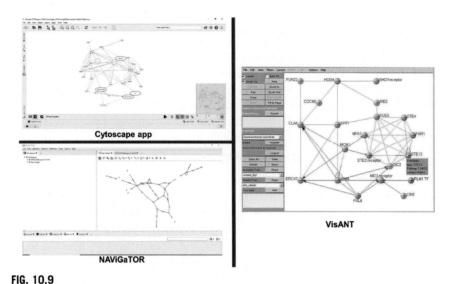

FIG. 10.9

Graphical Unser interface of Cytoscape app, NAVIGaTOR and VisANT.

graphs in both 2D and 3D are available and we can combine all nodes into one "meta node". NAViGaTOR also contains API for plugins and different data formats. Lasso selection method and bookmarking to support manual design and other network analysis operations (Fig. 10.9).

10.11.3 **VisANT**

VisANT (Hu et al., 2008) is a network visualization tool which can run in the browser or as a separate program. One of its interesting features is the name resolution feature, which tries mapping nodes in the network to different gene names, resulting in each protein coded by a gene to be depicted as a separate entity. This name-mapping feature is one of the easiest to use and understand as compared to many of the other software packages available. VisANT has been used on large datasets having a large number of nodes. Metagraphs can also be represented using this tool. In metagraphs, a single node can contain a subgraph. This tool is also integrated with online databases, containing a large number of interactions (Fig. 10.9).

10.11.4 **CellDesigner**

This is diagram editor tool used for drawing biochemical networks and gene regulatory networks. The user can search or alter networks in the form of process diagrams and stock the networks in standard for depicting these networks, known as systems biology markup language. This editor can connect the network with simulations. Users can see the network dynamics through a GUI (Funahashi et al., 2007). This editor is implemented in Java and integrated different packages.

10.11.5 **Pathway Studio**

Pathway Studio (Nikitin et al., 2003) is another free tool that uses NLP- based data extraction for gene-gene and protein—protein interactions. Pathway Studio contains databases of cellular pathways and protein interactions. This tool effectively scans biological words and terms occurring along with them. The database has a large library of interactions and proteins extracted from literature present on different organisms.

10.11.6 **Gephi**

A program that can work on large networks without the requirement of programming skills is Gephi ("Gephi - The Open Graph Viz Platform," n.d.). Gephi can handle a large number of nodes and edges. However, this software requires a large amount of computational power. The advantages of using Gephi are that this software is free, can be used on different platforms and can use advanced algorithms as plugins. The limitation of this software is that it lacks the ability to analyze specific biological data. This tool can be used for visualizations, statistical analysis and enumerating.

10.12 **Important properties to be inferred from networks**

The first step of all network analysis processes is the construction of a network. Investigation of features in a network and their deviation from our expectations helps in better understanding the networks. Different computational and mathematical methods have been generated for the analysis of large networks for identification of the selected features.

10.12.1 **Hubs**

Node degree is an informative feature in network analysis. Interaction networks are usually present in a scale-free topology. In these networks, hub nodes are important for the functions and structure of these networks.

10.12.2 **Bottlenecks**

Nodes with high betweenness nodes are known as bottleneck nodes. This means, that these nodes are present in paths between many different nodes. Bottlenecks play an important role in the communication in a network, as they help in the flow of information among modules (Sub networks with dense connections). These nodes in a network are present on the major connections. Disturbances in the bottleneck nodes can lead the network into a disarray, due to presence of a low number of connections near bottlenecks. Bottleneck nodes are more important to networks as compared to the hub nodes, causing these bottleneck nodes to be targeted by pathogens.

10.12.3 **Modules**

Networks are modular and have communities in their structure. Specific modules are usually enriched to perform the usual biological functions. Therefore, detecting modules present in networks can help understand processes/functions that are not known in a pathway. Proteins related to a disease or disease having same phenotypes usually display interactions in modules of the disease. Therefore, we can detect modules that show enrichment in genes or proteins linked to a disease. Other module proteins which are not known to be linked to a disease, can be predicted to be associated to the disease. This method is now popularly being used to detect disease modules for a variety of diseases in humans. E.g., A network module controlling the heart development has been associated with heart disease (Fig. 10.10).

10.12.4 **Bioinformatics tools to detect modules, bottlenecks and hubs**

A broad range of tools are currently used for the fast detection of bottlenecks and hubs. An example is NetworkAnalyst, which can analyze networks based on gene expression. NeworkAnalyst can take a list of genes given by the user and interactions involved from InnateDB to find modules, betweenness centralities and degrees in the network. Cytoscape also contains applications that can help in network analyses. An example of such application is cytoHubba, which can detect bottlenecks and hubs in the network given as input to cytoscape. Cytoscape can be utilized along with the networks produced InnateDB. A wide range of tools have been created to detect network modules.

NetworkAnalyst has many analytical features that can detect network modules. NetworkAnalyst utilizes random walk algorithm to detect modules of high frequency nodes. This tool can build an edge weighted network, where node information like gene expression can be used to give weights. Cytoscape has many different applications for detection of modules. One such application is jActiveModules which detects connected parts of the network that have significant differences in

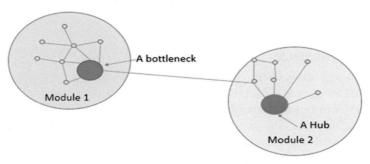

FIG. 10.10

Figure depicting hub, bottleneck and modules.

expression of genes. When modules associated with diseases need to be identified, other algorithms give a better performance, since proteins associated with diseases are not present in dense regions and nodes associated with diseases may be identified more accurately through connectivity significance. The DIAMOnD (Disease Module Detection) algorithm is a novel method to find disease modules on the basis of connectivity significance. Incompleteness of the interactome also restricts the disease modules which can be identified, and the small quantity of known disease linked proteins which can be used to detect the disease.

References

Altaf-Ul-Amin, M., Afendi, F.M., Kiboi, S.K., Kanaya, S., 2014. Systems biology in the context of big data and networks. BioMed Res. Int. 2014. https://doi.org/10.1155/2014/428570.

Aranda, B., Blankenburg, H., Kerrien, S., Brinkman, F.S.L., Ceol, A., Chautard, E., Dana, J.M., de Las Rivas, J., Dumousseau, M., Galeota, E., Gaulton, A., Goll, J., Hancock, R.E.W., Isserlin, R., Jimenez, R.C., Kerssemakers, J., Khadake, J., Lynn, D.J., Michaut, M., O'Kelly, G., Ono, K., Orchard, S., Prieto, C., Razick, S., Rigina, O., Salwinski, L., Simonovic, M., Velankar, S., Winter, A., Wu, G., Bader, G.D., Cesareni, G., Donaldson, I.M., Eisenberg, D., Kleywegt, G.J., Overington, J., Ricard-Blum, S., Tyers, M., Albrecht, M., Hermjakob, H., 2011. PSIC-QUIC and PSISCORE: Accessing and scoring molecular interactions. Nat. Methods 8 (7), 528—529. https://doi.org/10.1038/nmeth.1637.

Barabási, A.L., Oltvai, Z.N., 2004. Network biology: Understanding the cell's functional organization. Nat. Rev. Genet. 5 (2), 101—113. https://doi.org/10.1038/nrg1272.

Breuer, K., Foroushani, A.K., Laird, M.R., Chen, C., Sribnaia, A., Lo, R., Winsor, G.L., Hancock, R.E.W., Brinkman, F.S.L., Lynn, D.J., 2013. InnateDB: Systems biology of innate immunity and beyond—recent updates and continuing curation. Nucleic Acids Res. 41, D1228. https://doi.org/10.1093/NAR/GKS1147.

Brown, K.R., Otasek, D., Ali, M., McGuffin, M.J., Xie, W., Devani, B., van Toch, I.L., Jurisica, I., 2009. NAViGaTOR: Network analysis, visualization and graphing Toronto. Bioinformatics 25, 3327—3329. https://doi.org/10.1093/BIOINFORMATICS/BTP595.

Chen, E.Y., Tan, C.M., Kou, Y., Duan, Q., Wang, Z., Meirelles, G.V., Clark, N.R., Ma'ayan, A., 2013. Enrichr: Interactive and collaborative HTML5 gene list enrichment analysis tool. BMC Bioinf. 14, 1—14. https://doi.org/10.1186/1471-2105-14-128/FIG-URES/3.

Frias, S., Bryan, K., Brinkman, F.S.L., Lynn, D.J., 2015. CerebralWeb: A Cytoscape.js plug-in to visualize networks stratified by subcellular localization. Database 41. https://doi.org/10.1093/DATABASE/BAV041.

Funahashi, A., Morohashi, M., Matsuoka, Y., Jouraku, A., Kitano, H., 2007. CellDesigner: A graphical biological network editor and workbench interfacing simulator. In: Introduction to Systems Biology, pp. 422—434. https://doi.org/10.1007/978-1-59745-531-2_21/COVER.

Gephi - The Open Graph Viz Platform [WWW Document], (n.d.). URL https://gephi.org/ (accessed 12.17.2022).

Hu, Z., Snitkin, E.S., Delisi, C., 2008. VisANT: An integrative framework for networks in systems biology. Briefings Bioinf. 9, 317. https://doi.org/10.1093/BIB/BBN020.

Huang, D.W., Sherman, B.T., Tan, Q., Kir, J., Liu, D., Bryant, D., Guo, Y., Stephens, R., Baseler, M.W., Lane, H.C., Lempicki, R.A., 2007. DAVID bioinformatics resources: Expanded annotation database and novel algorithms to better extract biology from large gene lists. Nucleic Acids Res. 35. https://doi.org/10.1093/nar/gkm415.

Kovács, I.A., Luck, K., Spirohn, K., Wang, Y., Pollis, C., Schlabach, S., Bian, W., Kim, D.K., Kishore, N., Hao, T., Calderwood, M.A., Vidal, M., Barabási, A.L., 2019. Network-based prediction of protein interactions. Nat. Commun. 10 (1), 1−8. https://doi.org/10.1038/s41467-019-09177-y.

Medina, I., Carbonell, J., Pulido, L., Madeira, S.C., Goetz, S., Conesa, A., Tárraga, J., Pascual-Montano, A., Nogales-Cadenas, R., Santoyo, J., García, F., Marbà, M., Montaner, D., Dopazo, J., 2010. Babelomics: An integrative platform for the analysis of transcriptomics, proteomics and genomic data with advanced functional profiling. Nucleic Acids Res. 38, W210. https://doi.org/10.1093/NAR/GKQ388.

Nikitin, A., Egorov, S., Daraselia, N., Mazo, I., 2003. Pathway studio—the analysis and navigation of molecular networks. Bioinformatics 19, 2155−2157. https://doi.org/10.1093/BIOINFORMATICS/BTG290.

O'Connor, S.E., 1992. Network theory—a systematic method for literature review. Nurse Educ. Today 12, 44−50. https://doi.org/10.1016/0260-6917(92)90009-D.

Shannon, P., Markiel, A., Ozier, O., Baliga, N.S., Wang, J.T., Ramage, D., Amin, N., Schwikowski, B., Ideker, T., 2003. Cytoscape: A software environment for integrated models of biomolecular interaction networks. Genome Res. 13, 2498. https://doi.org/10.1101/GR.1239303.

Thomas, P.D., Ebert, D., Muruganujan, A., Mushayahama, T., Albou, L.P., Mi, H., 2022. PANTHER: Making genome-scale phylogenetics accessible to all. Protein Sci. 31, 8−22. https://doi.org/10.1002/PRO.4218.

Yon Rhee, S., Wood, V., Dolinski, K., Draghici, S., 2008. Use and misuse of the gene ontology annotations. Nat. Rev. Genet. 9 (7), 509−515. https://doi.org/10.1038/nrg2363.

Bioinformatics workflow management systems

11

11.1 Introduction to workflow management systems

In this era of data-driven science, the application of bioinformatics is undergoing a fundamental transformation. Similar to the development of cosmology and the physical sciences, genomics has grown to become a significant information science. It is accompanied by a shift away from local high-performance computing (HPC) facilities and toward circulated networks and, more recently, cloud resources, particularly in large-scale multi-focus cooperative projects. In a similar way, the need to process more data at a faster rate is driving the development of software that can automate and speed up the process of looking at data in high-performance computing environments (Shade and Teal, 2015).

Scientific Workflow Management Systems, often known as WfMSs, are capable of automating computational studies by merging a variety of data-processing procedures into a single pipeline. They hide problems associated with coordinating the processing and transportation of data, managing task dependencies, and assigning resources within the computational infrastructure. There are additional WfMSs that provide techniques for monitoring the provenance of data, analyzing execution faults, authenticating users, and ensuring data security. The development of the Findable, Accessible, Interoperable, and Reproducible (FAIR) principles for scientific tools, workflows, and protocols for sharing datasets was prompted by the growing use of workflow management systems (WfMSs) in contemporary scientific research. Currently, containerized software and standard Application Programming Interfaces (APIs) are being built based on these characteristics in order to facilitate the creation, distribution, and execution of code in a variety of computer settings (Perkel, 2019).

The motivation for the development of a WfMS may have an effect on its ease of use, the functionality and features it provides, and how effectively it operates. The Common Workflow Language, often known as CWL, is a language specification that was established by the community of bioinformatics to unify the style, principles, and standards of coding pipelines in a way that was independent of the hardware being used. The repeatability of workflows and their portability are at the very top of this system's priority list. As a consequence of this, it is necessary to provide detailed and exact explanations of the parameters, which results in a long paper. Workflow

Description Language, often known as WDL, is a definition of a language that places an emphasis on making code easy for humans to comprehend and learn, but at the price of being less expressive. The combination of a workflow language and an execution engine is at the heart of the comprehensive solution known as Nextflow. It's possible that this WfMS is one of the most developed ones ever made. Nextflow has features such as easy reading, a compact size, agile, and the ability to track where the data came from. On the other hand, coding is quite easy to understand, especially for people who are just beginner at biological computing. The Swift/T system is an all-encompassing solution in the same way. Swift is the programming language that is used by Turbine, which is the execution engine. It was designed by engineers and physicists with the intention of providing short, rapid-fire tasks at exascale in a scalable way. As a consequence of this, it is a very low-level language (similar to C) that is very powerful yet challenging to master. In the following sections, we will explain how these distinct ways of thinking influence the use of the four WfMS that may be used in production bioinformatics (Leipzig, 2017).

The most important parts of a typical cloud WFMS are shown in Fig. 11.1. Creating and describing abstract workflows made up of activities and the interdependencies between them may be done using the workflow site (Ahmed et al., 2021). Reading of the abstract workflows is handled by the language parser, which is a component of the workflow enactment engine. After that, the task dispatcher investigates the connections between the different jobs and sends the finished work to the scheduler for processing. The scheduler selects a resource to do the workflow task based on the scheduling techniques that have been provided. In the next part, we will go further into the topic of task scheduling and cover it in more depth. The fault tolerance of the process is taken care of by the enactment engine for the workflow. In addition to that, it has a function for allocating resources to tasks. This is made possible through the use of the resource broker.

The resource broker is in charge of communication with the infrastructure layer, and it gives the enactment engine a comprehensive overview of everything that's going on. In order to get the necessary resources, the resource broker will interface with the computing services. Information on data objects, programs, and computer resources is stored by the directory and catalog services. This information will be used by both the enactment engine and the resource broker to come to important conclusions.

In general, workflow management services provide essential activities that are necessary for the operation of a workflow management system (WfMS). The authentication process and safe access to the WfMS are both provided by the security and identity services. Monitoring tools keep a watch on the WFMS's most important components and provide warnings whenever they deem it essential to do so. The database management component offers safe storage space for both the intermediate and final data outcomes produced by procedures. Execution information, file locations, input and output information, workflow structure, form, workflow evolution, and system information are some of the important pieces of data that are recorded by provenance management systems. Other types of data that are recorded include

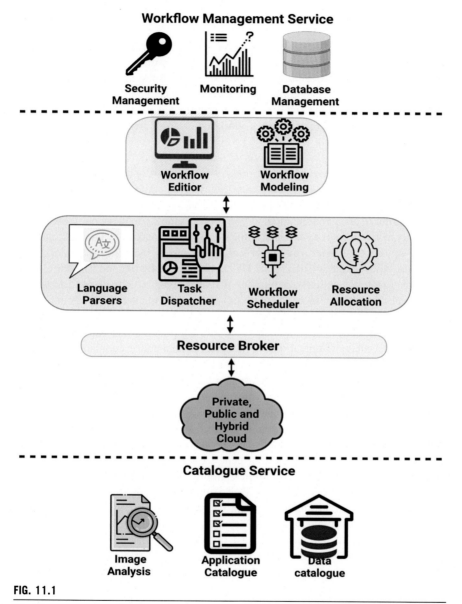

FIG. 11.1

Workflow management system.

the dynamics of control flows and data as well as how they change. Understanding the data, figuring out its quality, defining who owns it, generating repeatable outputs, maximizing efficiency, addressing concerns, and making allowances for faults are all activities that need provenance.

11.2 Galaxy

Galaxy (Cock et al., 2013) is a platform for scientific workflows that enables the integration of data, the storage of data, and analyses, as well as publishing. Research scientists who are not acquainted with programming or system administration will be able to use computational biology thanks to this project's goal of making it accessible to them. In spite of the fact that it had its beginnings in genomics research, it is today being used as a bioinformatics workflow management system for several different sorts of research.

Functionality: The Galaxy system is a way for researchers to structure their research and data. These technologies make it possible to conduct computer studies in stages, much like the steps in a recipe. They often come with a graphical user interface that gives you the ability to choose the data you want to work with, the actions you want to take, and the method you want to employ to carry out those tasks.

Galaxy also functions as a database for storing biological information. It enables users to upload data from a range of online resources, including their desktops, URLs, and other websites (such as the UCSC Genome Browser, and BioMart). Galaxy is able to interact with a wide number of conventional biological data formats as well as translate between them. Academics can format and alter text without having to learn how to code, thanks to Galaxy, which gives a web interface to a variety of text manipulation tools. Galaxy's capability for manipulating intervals may be used to carry out set-theoretic operations on the intervals that are being manipulated. The name of a chromosome or contig, as well as its beginning and ending coordinates, are examples of the types of genomic interval data that are included in a great number of biological file formats. This makes it possible to combine these pieces of data.

Galaxy was developed for the purpose of analyzing biological data, namely genomics. Over the course of time, there has been a massive expansion in the selection of tools that are at a user's disposal. Galaxy is presently used for research into a wide range of life science topics, including gene expression, genome assembly, proteomics, epigenomics, and transcriptomics, among others. Because the platform is not exclusive to any one scientific field, it is open to use in any and all scientific fields, including cheminformatics. Image classification, combinatorial chemistry, targeted therapies, cosmology, climate science, social research, and linguistics are just a few of the fields that may make use of the servers that Galaxy provides.

The Galaxy Tool Shed (https://usegalaxy.org/toolshed) is a central area that allows tool authors to exchange their tool settings as well as "recipes" for installing dependencies with other tool writers. This makes it easier for different Galaxy instances to share tools with each other.

Interactive analysis and visualization: Interactive analysis and visualization are two of the features offered by the Galaxy user interface, which makes it possible for anybody to do complex research. However, in order to conduct a comprehensive study of genetic data, it is often necessary to create custom scripts or visualizations, especially at the beginning (data preparation) or the end (data analysis) (data summarizing). Galaxy Interactive Environments, a Galaxy interface with Jupyter

(RStudio is in progress), a widely used interactive programming environment, was recently provided in order to meet these particular criteria. This was done so that Galaxy could fulfill these requirements. Galaxy users can make use of their current computer infrastructure by using either graphical user interfaces or ad hoc programming, or any combination of the two.

Community: As a consequence of contributions from the community, both the Galaxy framework and the collection of tools have undergone substantial development. Over the course of the last 2 years, 174 engineers have contributed to the effort to make Galaxy more scalable, functional, and user-friendly. Because of this, there have been 13,135 contributions, which is a 63% increase compared to January 2016. The project makes use of the Travis and Jenkins continuous integration (CI) services in order to do thorough automated testing on each proposed set of code modifications. This approach lowers the total number of bugs that are introduced into the codebase while also speeding up the review process.

By using the open-source community and the conventional approach to software development, the foundation is capable of releasing a stable version of the Galaxy framework every 4 months. Some of the current directions for the future development includes the following: toolshed installation and development upgrades; data and compute federation; tighter coupling of Interactive Environments with provenance and reuse; continued development of collections, processes, analysis interfaces, and historical views; more training materials; better statistical usage tracking and instrumentation; and a lot more.

Anyone who is interested in contributing to Galaxy is encouraged to read the project's contributing and code of conduct guidelines, search for outstanding issues, and review the current roadmap, all of which are available on the Galaxy GitHub repository. Galaxy has been of assistance to tens of thousands of people on a daily basis, and it has been cited in more than 5700 scientific works. In addition to this, it has provided over 500 developers with a framework that makes the process of data analysis straightforward, observable, and re-useable. Galaxy Main (and more than 99 other public Galaxy servers) have been set up in order to make it easier to conduct research in many fields, including biological research and study in other domains. During the last 2 years, the Galaxy Project has made great headway in all fields, which has led to overall improvement (Fig. 11.2).

11.3 Gene pattern

GenePattern (Kuehn et al., 2008) is a piece of computational biology software that was created at the Broad Institute and is available for free and open-source use. It is used for the processing the genomic data. GenePattern was originally made available to the public in 2004, and its primary purpose was to provide academics with a platform on which they could construct, store, and implement genomic analysis algorithms. GenePattern is currently being developed by researchers at the University of California, San Diego.

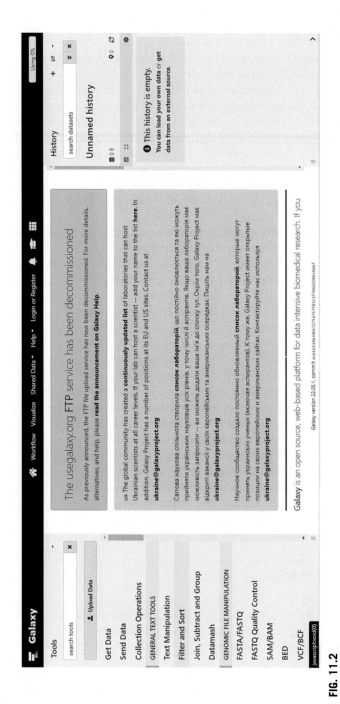

FIG. 11.2

Galaxy interface.

GenePattern is a repository of hundreds of bioinformatics analysis and visualization techniques (referred to as "modules"). It also contains utilities for data formatting, preprocessing, and other functions that serve as "glue" between analysis steps. Simply pointing and clicking will get you through the user interface; no coding is required. Since 2008, individuals have been authorized to use the public GenePattern server, which can be found at www.genepattern.org. It has a user base of around 40,000 people and does between 2000 and 5000 analytical activities each week. Public servers are also accessible at Indiana University and the Garvan Institute. More than 17,000 research groups, bioinformatics core centers, and individual scientists have downloaded and used the software.

GenePattern is software that operates on both a client and server level. Components of an application may be operated on a single computer with requirements as low as those of a laptop, or they can be distributed over several computers, allowing the server to make use of hardware with a higher processing capacity. The GenePattern server is really the GenePattern engine. It runs the analysis modules and stores the results of those analyses when they are generated. Both the Web Client and the Desktop Client are examples of graphical user interfaces that simplify the process of accessing the server and all of the modules that it contains. A web browser is required in order to access the Web Client once it has been installed concurrently with the server. The desktop client is an independent piece of software that must be downloaded and installed on each individual machine. GenePattern libraries for the Java, MATLAB, and R programming environments allow you to access the server and its modules via function calls in addition to providing access to those environments. The essential protocols described in this part are carried out with the help of the Web client. However, these protocols may also be carried out with the assistance of the desktop client or a programming environment (Fig. 11.3).

The majority of transcription profiling studies aim to accomplish at least one of the following goals: investigate differences in gene expression; find new classes; or anticipate future classes. The goal of differential expression analysis is to discover (if any) genes that are expressed differently in distinct groups or kinds of samples. This may be done by comparing the gene expression levels of a number of different groups or types of samples. Genes that express themselves in unique ways are referred to as marker genes, and the process that identifies these genes is referred to as marker selection. The process of merging genes or samples that have similar expression profiles into a smaller number of patterns or "classes" is known as class discovery. This process offers a high-level view of the microarray data. It is possible to identify common biological processes by categorizing genes according to the degree to which their expression patterns are comparable to one another. It is possible to identify common biological states or subtypes of diseases by grouping samples according to the degree to which their expression patterns are comparable to one another. There are a number of different ways that one may use data on gene expression to find classes. The goal of research into class prediction is to locate crucial marker genes whose patterns of expression can correctly categorize unlabeled data.

FIG. 11.3

GenePattern interface.

- Accessibility: Utilize a point-and-click user interface to run over 200 analysis and visualization tools, such as data preprocessing, gene expression analysis, proteomics, single nucleotide polymorphism (SNP) analysis, flow cytometry, and next-generation sequencing, and develop analytical workflows without the need for scripting.
- Reproducibility: Any user is able to understand, duplicate, and share the results of a whole computational research thanks to versioning, automated history tracking, and provenance monitoring.
- Extensibility: The computational approach, users are able to collaborate and share their processes and code with one another by using tools that simplify the generation and combination of such elements.

11.4 KNIME: The Konstanz information miner

The KNIME Analytics Platform (Fillbrunn et al., 2017) is open-source software for data science, which means that anybody is free to use it. Anyone may do data analysis and construct data science processes and reusable components with the help of KNIME since it is user-friendly, open-source, and continually updated with new features. The following is a list of some of the features that are included in KNIME Builds Workflows: Using a simple drag-and-drop user interface, it is possible to create visual workflows even if you lack writing skills. The user has the ability to construct workflows, model each step of an inquiry, manage how data flows, and guarantee that the work is up-to-date by picking from over 2000 nodes that are provided by KNIME. The software also incorporates other technologies, such as machine learning, Python or R programming, or Apache Spark connections, into a single workflow. This is done by integrating KNIME native nodes with these extra technologies (Fig. 11.4).

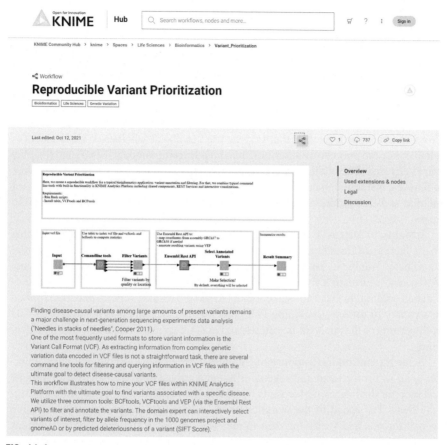

FIG. 11.4

Knime variant prioritization workflow.

Data that is combined: KNIME allows for the mixing of a variety of data kinds and formats, including structured and unstructured data types, as well as time-series data. Some examples of these formats are PDF, CSV, XLS, and JSON. It establishes connections to Oracle, Microsoft SQL, Apache Hive, and other databases and data warehouses so that it may aggregate data from those sources. Data comes from many different places, such as Twitter, AWS S3, Google Sheets, and Azure.

Data on Shapes: performs calculations using statistics such as the mean, quantiles, and standard deviation, or uses statistical tests in order to establish a hypothesis. Workflows could also incorporate other elements like correlation analysis, cutting down on the number of dimensions, and other similar things. Data can be collected, sorted, filtered, and combined on a single workstation, in a database, or in a "big data" environment where the data is spread out over several workstations and databases.

Machine Learning and Artificial Intelligence: provides machine learning models for classification, regression, dimension reduction, or clustering utilizing advanced methodologies such as deep learning, tree-based algorithms, and logistic regression. These models may be used to analyze large amounts of data. Increasing the performance of a model may be accomplished in a number of different ways, such as by adjusting the hyperparameters, boosting, bagging, stacking, or generating sophisticated ensembles. Performance measures like accuracy, R2, the area under the curve (AUC), and ROC are used to determine how well a model functions. Cross-validation is a method that may be used to check whether or not a model is reliable. For the purpose of describing machine learning models, LIME and Shapley values are used. The interactive presentation of partial dependency and ICE helps the tool have a better understanding of the model's predictions and make forecasts based on tried-and-true models or the most advanced PMML available, including Apache Spark.

Insights to Discover and Share: The data is presented via user-customizable charts that range from straightforward (bar charts, scatter plots) to intricate (parallel coordinates, sunburst, network graph, heat map). This command will remove any extraneous information from a KNIME table and show a summary of the statistics for each column. In order for stakeholders to see the results, reports may be exported in a variety of formats, including PDF, PowerPoint, and others. The processed data or the results of the analysis are often saved in a file format or database that is widely used.

Scale Execution in Response to Demands: develops workflow prototypes in order to allow for the testing of several different approaches to data analysis. Examine and monitor the intermediate results in order to get rapid feedback and discover original ideas. The speed of the workflow may be increased by processing data using several threads and streaming data in memory. Increases the speed of calculations by using either the database processing capabilities of Apache Spark or its distributed computing capabilities.

KNIME Extensions provide access to strong machine learning algorithms as well as a diverse selection of complex data types. KNIME Analytics Platform and KNIME Server collaborate with many other open-source projects. Access data from Apache Hadoop and systems that store Hadoop data, such as Hive and Impala. You may model and execute Apache Spark jobs in local KNIME configurations to leverage the potential of scalable analytics. Create a model for generating predictions, then apply it to fresh data, or just use R or Python code inside a KNIME process to create various forms of visualization. Deep neural networks may be read, created, edited, trained, and executed.

HiLiting: KNIME is strongly reliant on the HiLiting system. It gives the user the ability to choose and highlight a number of rows inside a data table, and those rows are then highlighted in all of the other views that show the same data table (or at least the high-lighted rows), Due to the one-to-one relationship that exists between the tables' distinct row keys, it is not difficult to implement this sort of highlighting. There are numerous nodes that change the structure of the input table, but the rows that are linked between the input table and the output table do not change. The techniques used for clustering data are an excellent illustration of a 1:n link. One of the inputs that the node receives is the training (or test) patterns, and one of the outputs that it produces is the cluster prototypes. Each cluster is in charge of its own unique collection of pattern inputs. When one or more clusters in the output table are highlighted, any input patterns in the input table that relate to those highlighted clusters are also highlighted. It is possible that other summarizing models, such as the branches and leaves of a decision tree, common patterns, and discriminative molecular fragments, may be translated in the same way.

11.5 **LINCS tools**

The LINCS project (Xie et al., 2022) was established on the idea that if any one of the several steps of a biological process is altered, the molecular and cellular features, behavior, and/or function of the cell will be altered. This idea serves as the project's guiding principle. The term for this characteristic is "cellular phenotype." By watching how and when different inputs change the phenotype of a cell, scientists may be able to understand more about the underlying mechanisms that lead to chaos and, ultimately, illness.

A number of data releases were be carried out in order to make the LINCS data available to the general public as a community resource. Because of this, scientists are able to work on a wide array of basic research problems, which will make it easier to discover biological targets for innovative disease treatments. The LINCS databases include the findings of tests conducted on cultured human cells as well as primary human cells that were treated with bioactive small compounds, ligands such as growth factors and cytokines, or genetic changes. These tests were performed on human cells. Assays that measure transcript and protein expression, in addition to biochemical and imaging readouts that collect data on cell phenotypic characteristics, are used so that the response of cells may be monitored.

The LINCS project is a two-part study funded by the NIH Common Fund. 2013 was the last year of the pilot phase of the program, which was made up of the following parts:

- Large-scale signatures of the molecular and cellular modifications that were produced by the disruption.
- In order to provide users with access to data, databases, data standards, and public user interfaces are now being developed.
- Developing new computer programs and gathering information.
- Recent advances in inexpensive molecular and cellular phenotyping

11.5.1 **The program's overall goal**

By employing an integrative strategy that searches for patterns of common networks and cellular responses (termed "cellular signatures") across various types of tissues and cells in response to a variety of changes, LINCS is attempting to find a novel approach to understanding health and illness. This is being done in order to advance the mission of the organization. The LINCS program is established on the idea that changing any one of the many stages that make up a biological process may result in changes to the molecular and cellular characteristics, behavior, and/or function of a cell. This concept serves as the program's guiding principle. After that, the phenotypic state of a cell might be represented by signatures that were acquired through similar clinical tests. By examining how and when different stimuli change the phenotypic characteristics of a cell, scientists may be able to learn more about the systems that are involved in disruption and, ultimately, sickness.

A collection of data descriptors with decreased dimensionality that may both shed light on a process and serve as a basis for making predictions is what is meant to be understood as a cellular signature of a response to a perturbagen. As a consequence of this, significant fingerprints are established based on the assay and the manner in which a number of tests are combined, either into prediction patterns or signaling networks that may lead to mechanistic explanations. The results of many tests need to be normalized and compared before they can be considered relevant. Integration, normalization, and scaling of heterogeneous, multi-parameter dose-response data is a difficult process that requires both theoretical and practical considerations.

11.5.2 Test performed under LINCS

The strategy calls for the establishment of six Data and Signature Generation Centers. The Drug Toxicity Signature Generation Center is comprised of a number of different centers, including the HMS LINCS Center, the LINCS Center for Transcriptomics, the LINCS Proteomic Characterization Center for Signaling and Epigenetics, the MEP LINCS Center, and the NeuroLINCS Center, among others. The Drug Toxicity Signature Generation Center developed assays that measure gene and protein expression in addition to phenotypic characteristics in order to gain a better understanding of how differentiated iPSCs react to single and multiple FDA-approved medication modifications. This was done in order to learn more about how differentiated iPSCs respond to the changes.

The HMS LINCS Center employs a number of different biochemical, imaging, and cell biology tests in order to track the responses of cells. Imaging experiments, transcriptional response tests (done in partnership with the LINCS Center for Transcriptomics), cell viability assays, and direct drug-kinase interactions in cell extracts are some of the others.

The HMS LINCS Center employs a number of different biochemical, imaging, and cell biology tests in order to track the responses of cells. Imaging experiments, transcriptional response tests (done in partnership with the LINCS Center for Transcriptomics), cell viability assays, and direct drug-kinase interactions in cell extracts are some of the others (Fig. 11.5).

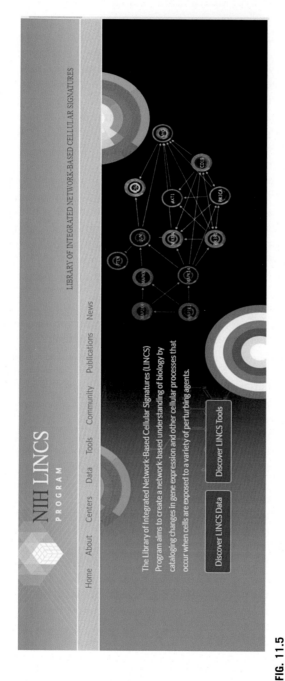

FIG. 11.5

Lincs interface.

11.6 Anduril bioinformatics and image analysis

Anduril (Cervera et al., 2019) is a tool for doing process analysis on massive amounts of data. It comes with ready-to-use tools for conducting analyses in the field of molecular biology, which may also be put to use for a variety of other tasks. Anduril makes it easy for people who use analytics to see the results, and it gives analysts a strong workflow environment in which to work.

The iterative workflow engine is used to run workflows in parallel once they have been created using Scala 2.11, which is used to generate the workflows. External libraries (like R and Python) and tools run from the command line are sometimes included in workflows. Anduril may be set up on a single computer or over an entire network of Linux computers and devices.

Software Development: The vast majority of the most famous bioinformatics frameworks, such as Anduril 2, are able to handle serial and parallel processes, intricate dependencies, a wide range of software and data file formats, as well as user-defined parameters and deliverables. Anduril presents the connections between components in the form of a graph and may automatically combine elements that do not depend on one another in any way. Because of the universal prefixing of the processes, SLURM and Sun Grid Engine may be used in a wide variety of different ways.

Re-entrancy: When running lengthy and complicated pipelines on enormous datasets, it is to the user's advantage to be able to resume execution from the point where it was stopped. This prevents the user from having to figure out where to start again or which samples have already been processed, which is a time-consuming and laborious task. You are free to make changes to the component settings or add new samples to the process at any point without the unfinished independent stages being forced to run. As a direct result of this, much less time is needed for both the calculation and the programming.

Dependency support: Any change to a step, such as an adjustment to a parameter, will cause all of the processes that are dependent on that step to begin over from the beginning. It is possible to annotate the inputs and outputs of a component in order to create artificial connections between them in cases where they are not physically linked. A component, for example, may not have any outputs, but it could still be able to change its environment in some way, such as by altering a database, which would then trigger the execution of another component on which it depends.

Resources in bioinformatics: There are around 400 well-described parts and functions that may be used to carry out standard tasks for a variety of bioinformatics research projects. These components can be used. The bulk of third-party software that is supported by Anduril components comes with its own installer, which makes it a great deal easier to install the many software packages that are necessary for standard bioinformatics research. Anduril 2 may utilize either its own installation or one that the user has specified on their end. Additionally, since the effective configuration of each component is stored in a bash script, any component may be operated outside of Anduril 2 with the same settings and pipeline inputs, provided it has the same script. Testing is made easier as a result.

Ease of integrating new tools or custom analysis: It is not difficult to include additional tools in a pipeline since eval-based components may be integrated with either their own code or the code of a third party. To add a new tool to the component repository for either private or public use, all you need to do is define its inputs, parameters, and outputs in an XML file, preferably together with test cases. After that, the tool will be available for use. For example, Taverna necessitates the use of third-party software in order to develop plug-ins that may be integrated into pipelines. Both Galaxy 2 and Anduril 2 make it easy to develop wrappers; however, Anduril also makes it easy to incorporate custom analysis and software into any pipeline. Galaxy 2 is the more recent of the two.

11.6.1 Anduril image analysis: ANIMA

The modern microscope generates an enormous quantity of picture data, which then has to be analyzed and interpreted using techniques specific to computers. In addition, a single image analysis project may need tens or hundreds of steps of analysis, beginning with the entry of data and pre-processing, moving on to segmentation and statistical analysis, and concluding with the visualization and reporting of the results. Anima is a workflow framework that is modular in nature and is used for organizing large-scale picture data processing projects.

Digital Image Analysis: The use of high-tech equipment such as X-ray machines, microscopes, and MRI and CT imaging equipment to produce digital photographs has been an integral part of our day-to-day lives for quite some time. This includes the taking of digital photographs with smartphones and their subsequent sharing on social media. The production and transmission of semantic information from digital images are referred to as "digital image analysis," and it is a technique that is in great demand across a variety of industries and applications.

The Anima workflow system, which is an acronym that stands for ANduril IMage Analysis, is a modular system that gives us the ability to analyze images in a quick and comprehensive manner. By enabling batch processing, Anima has made high-throughput image analysis very easier to do.

Processing, interoperability, and the freedom to run on any platform are just some of the perks that come along with this. Anima is mainly geared toward developers of algorithmic and analytic tools, and it gives these developers the ability to combine a wide variety of computing tools into a single workflow system. The primary objective of Anima is to make Rapid Application Development (RAD) possible as well as the incorporation of new methods without the need of porting them from their original implementations. Regardless of the computer language that is being used, Anima gives researchers in the field of image analysis the ability to experiment with novel methods and integrate their findings into established procedures.

Anima's architecture has a fundamental flaw in that it was designed to be used as a platform for enhancing the functionality of other apps rather than as a replacement for such programmes in their entirety. Anima has proven to be a useful tool for a variety of tasks, including the creation and evaluation of innovative algorithms as well as the execution of routine analyses. It has a horizontal data flow architecture, which means that each step, like picture segmentation, can be done at the same time on all of the images.

Horizontal flow management gives you the ability to make certain that each stage of processing yields the desired results before moving on to the next phase. Because Anima is a platform that is adaptable, extensible, and scalable, it is very easy to add new features to the system while also guaranteeing that those features may be used effectively several times. You are also able to quantify attributes based on all of the images using Anima, which you can then use in the process of analysis. Anima is a really powerful tool.

11.7 NextFlow

Nextflow (di Tommaso et al., 2017) is a reactive workflow architecture and domain-specific language (DSL) that was developed at the Barcelona Center for Genomic Regulation by the Comparative Bioinformatics Group (CRG). It lets scientists use software containers to make scientific procedures that can be scaled up and used over and over again.

Nextflow may be run either locally or on a dedicated instance of the Amazon Elastic Compute Cloud. If your procedures require a significant amount of computing resources, go with the second choice. However, after your procedure is over, you are required to terminate the instance. The architecture that is shown here explains how to make the most of AWS Batch in order to run Nextflow in a way that is both simple and economical to administer.

An RNA-Seq analysis may be carried out by following a sequence of steps known as the nextflow-core RNA-Seq workflow. This process was developed by the community. The DNA sequence data will first be converted into the raw FASTQ file format before undergoing quality control, alignment to the reference genome, quantification, and differential expression calculation. Over the course of its development, this pipeline has seen over 3700 separate changes, all of which can be seen on GitHub. The procedure, by default, consists of 20 steps and uses a number of different programs. When compared to starting from scratch, using this strategy will provide results far more quickly.

Workflows made with Nextflow should be able to handle millions of samples as long as enough computing resources are available. For instance, 23 and Me uses NextFlow to handle the genetic data of its customers. However, when biological data is converted into the structure and size that are generally handled by data engineering procedures, bioinformatics workflow managers may not be the best answer. Ginkgo Bioworks is a genomics company, whose pipeline processes terabytes of

FIG. 11.6

NextFlow interface.

sequencing data every single day. Ginkgo uses an Airflow, Celery, and Amazon Web Services batch process. At industry scale, efficiency is of the utmost importance, and the solution that Ginkgo has developed was developed by an entire team of data engineers. The majority of biotechnology companies as well as university labs would benefit more from using Nextflow or another bioinformatics-specific workflow management system that may be set up by a single scientist.

The capability of isolating the workflow implementation, which specifies the flow of data and the operations that are to be done on that data, from the configuration parameters that are needed by the execution platform is one of the most important aspects of Nextflow. This makes the process portable, so it can run on many different computer platforms, like an institution's high-performance computing (HPC) system or a cloud architecture, without having to change the way the workflow is designed (Fig. 11.6).

References

Ahmed, A.E., Allen, J.M., Bhat, T., Burra, P., Fliege, C.E., Hart, S.N., Heldenbrand, J.R., Hudson, M.E., Istanto, D.D., Kalmbach, M.T., Kapraun, G.D., Kendig, K.I., Kendzior, M.C., Klee, E.W., Mattson, N., Ross, C.A., Sharif, S.M., Venkatakrishnan, R., Fadlelmola, F.M., Mainzer, L.S., 2021. Design considerations for workflow management systems use in production genomics research and the clinic. Sci. Rep. 1 (11), 1−18. https://doi.org/10.1038/s41598-021-99288-8.

Cervera, A., Rantanen, V., Ovaska, K., Laakso, M., Nuñez-Fontarnau, J., Alkodsi, A., Casado, J., Facciotto, C., Häkkinen, A., Louhimo, R., Karinen, S., Zhang, K., Lavikka, K., Lyly, L., Pal Singh, M., Hautaniemi, S., 2019. Anduril 2: Upgraded large-scale data integration framework. Bioinformatics 35, 3815−3817. https://doi.org/10.1093/BIOINFORMATICS/BTZ133.

Cock, P.J.A., Grüning, B.A., Paszkiewicz, K., Pritchard, L., 2013. Galaxy tools and workflows for sequence analysis with applications in molecular plant pathology. PeerJ 1. https://doi.org/10.7717/PEERJ.167.

di Tommaso, P., Chatzou, M., Floden, E.W., Barja, P.P., Palumbo, E., Notredame, C., 2017. Nextflow enables reproducible computational workflows. Nat. Biotechnol. 35 (4), 316−319. https://doi.org/10.1038/nbt.3820.

Fillbrunn, A., Dietz, C., Pfeuffer, J., Rahn, R., Landrum, G.A., Berthold, M.R., 2017. KNIME for reproducible cross-domain analysis of life science data. J. Biotechnol. 261, 149−156. https://doi.org/10.1016/J.JBIOTEC.2017.07.028.

Kuehn, H., Liberzon, A., Reich, M., Mesirov, J.P., 2008. Using GenePattern for Gene Expression Analysis. Current Protocols in Bioinformatics/Editoral Board. Andreas D. Baxevanis ... [et al.] 0 7, Unit. https://doi.org/10.1002/0471250953.BI0712S22.

Leipzig, J., 2017. A review of bioinformatic pipeline frameworks. Briefings Bioinf. 18, 530−536. https://doi.org/10.1093/bib/bbw020.

Perkel, J.M., 2019. Workflow systems turn raw data into scientific knowledge. Nature 573, 149−150. https://doi.org/10.1038/d41586-019-02619-z.

Shade, A., Teal, T.K., 2015. Computing workflows for biologists: A roadmap. PLoS Biol. 13. https://doi.org/10.1371/journal.pbio.1002303.

Xie, Z., Kropiwnicki, E., Wojciechowicz, M.L., Jagodnik, K.M., Shu, I., Bailey, A., Clarke, D.J.B., Jeon, M., Evangelista, J.E., v. Kuleshov, M., Lachmann, A., Parigi, A.A., Sanchez, J.M., Jenkins, S.L., Ma'ayan, A., 2022. Getting started with LINCS datasets and tools. Curr. Protoc. 2, e487. https://doi.org/10.1002/CPZ1.487.

Data handling using Python

12

12.1 Introduction

Data handling is the process of collecting, storing and analyzing data. Data handling is a key skill for biologists, as understanding and processing data is integral to their work. Data handling is often associated with statistical analysis, which can be subjective or quantitative (Marx, 2013). Statistical analysis can often be used to answer questions about what qualifies as statistically significant and how large samples are necessary for accurate conclusions. Data handling begins with the data producer defining (1) what the type of data is (e.g., genetic, behavioral) and (2) the question to answer based on that type of data: for example, how many animals there are in a population on an island? The next step would be to choose an appropriate sampling method: random sampling, systematic sampling, or opportunistic sampling. The data producer then (3) gathers the data, records it in a spreadsheet or database, and then (4) analyzes and publishes their analysis. Data management is the process that ensures that raw and processed data can be organized efficiently in order to investigate specific questions related to the experimental task at hand (Hasija and Chakraborty, 2021).

In bioinformatics research, a variety of programming languages are often utilized, including Python, R, C, and Java. Python-based bioinformatics research is being introduced in this project. Guido van Rossum developed this all-purpose programming language, which was first made available in 1991. Python is a great language to learn for beginners into coding. Few basic characteristics of Python are:

Interpreted language: So, Python is a Interpreted Language as it itself is a software running on the computer. A Python script is just a text file when it is written. Python program may be instructed to do a variety of tasks on a text file, such as operations on genome's sequence or creating graphs etc. As opposed to compiled languages like C, which transforms the code into an executable. Generally speaking, compiled languages may operate more quickly than interpreted ones. The ability to quickly test every iteration of a script without having to do an additional compilation step, however, helps speed up the creation of new applications. A tradeoff between development time spent developing a code time the computer spends actually running your code. Because the development time is quite lengthy relative to the run time in many scientific applications where the user base is tiny and new code must be regularly produced, interpreted languages are often employed.

All About Bioinformatics. https://doi.org/10.1016/B978-0-443-15250-4.00008-3

High-level language: High level languages include a lot of built-in automated procedures that take care of various elements of how the code is written is used by a computer. For instance, Python takes care of allocating memory to keep the contents of the variable when it is defined, so user don't have to. In contrast, manual memory allocation is required for other languages like C. In general, writing a functioning program rapidly in a high-level language is substantially simpler. Low-level languages, on the other hand, may provide fine-grained control that can aid in the optimization of extremely computationally demanding applications.

Python has become one of the most popular programming languages in recent years. Its versatility and ease of use make it a great choice for a wide range of tasks, including data handling. Python has become the language of choice for data science and machine learning due to its simple syntax and powerful libraries. These libraries allow Python to be used for anything from web development to scientific computing. Python is especially popular in the field of data science. Some of the most popular Python libraries for data science are Pandas and NumPy. These libraries allow you to easily manipulate data and perform statistical operations (Ekmekci et al., 2016).

BeOpen Python Labs created the first version of Python 2.0, which was released in the year 2000. Prior to the formation of the team, Rossum (conceiver of Python) was in charge of the majority of Python's new features and bug fixes (van Rossum and Drake, 2009). He remained hopeful that Python would play a larger role in increasing computer science "literacy." Python was designed so that even someone with no prior knowledge of computers can use it. As a result, the Python Labs team created Python 2.X to make it less reliant on Rossum's management and more accessible to community contributions. Python 2.7 was the final and most recent version of Python 2. The Python 2 was discontinued in 2020. Even though Python 3.0 was released in 2008, it was more than just a bug-fixing version of Python 2. Instead, it introduced a significant change that rendered the language incompatible with subsequent editions. One purpose of the Python 3 syntax reform was to allow code to achieve the same thing in multiple ways ("Python 3.0 Release | Python.org," n.d.). The purpose of Python 3.X is to ensure that everyone utilizes the same, simple way. This eliminates the most common issues that novice programmers have when learning a new language. Python is a community-driven programming language that is widely used in the field of data science and machine learning. Because of its relatively flat learning curve, many experts recommend it as the first programming language to learn for beginners. Primarily, Python has an easy-to-understand English-like readable syntax.

12.2 Datatypes and operators

12.2.1 Datatypes

Data structures are specialized methods for storing and arranging information while programming. Each offers various methods of dealing with the data and is best suited for certain sorts of data. Gaining proficiency with Python's data types and understanding when to utilize each will help you develop into a professional Python programmer. Among the several datatypes, the four most prevalent are as follows:

- int (integers or whole numbers)
- float (decimal numbers or floating-point numbers)
- bool (Boolean or True/False)
- str (string or a collection of characters like a text)
- list, Tuple, set (Collection of Items

Python has two significant ways of representing numbers: int and float. Decimal values (floats) such as 1.0, 3.14, and −2.33 may take up more space than integers or whole numbers such as 1, 3, −4, and 0. Following that, there are Boolean datatype that is either "True" or "False"; these are utilized to create conditions, discussed in conditional statements. Finally, the "str" or string datatype is the most frequently used and encountered by biologists, as the majority of DNA, RNA, protein sequences, and names are text or strings. As a result, this chapter contains a distinct section on strings. It is critical to emphasize here that string data is always enclosed in quotes, i.e., ("<string data>"). For instance the peptide "MKSGSGGGSP" will be a Python string.

12.2.2 **Operators**

There are common operators as used like "+", "−", "*", "/", "=", and "**" for addition, subtraction, multiplication, division, assignment and exponent respectively.

Operation on an integer and a float will always produce a float type result, while operation on two integers will always return an integer type result, with the exception of division, which will always return a float type result. By employing the integral division operator, one can obtain an integer type in exchange for division. i.e., "//".

12.3 **Variables**

Variables in Python are analogous to algebraic variables in mathematics. Variables are composed of two components: a name and a value. Variables names declared and assigned aname by using the assignment "=" operator. On the left is the name, and on the right is the value (Table 12.1).

Variables can be recalled after they have been assigned. As seen in Table 12.2, variables "weight" in kilogrammes and "height" in meters are assigned and then utilized to calculate BMI, which is subsequently stored in another variable named "bmi."

Variables make our programmes readable and reusable. For example, if one is working with a large protein or nucleotide sequence, it would be imprudent to write it every time. As a result, it can be saved in a variable and reuse it whenever it is

Table 12.1 Assigning variables in Python.

Code	Output
`weight = 75` `print(weight)`	75

Table 12.2 Operations with variables.

Code	Output

`weight = 75`
`height = 1.5`
`bmi = weight/height`
`print(bmi)`
	50.0

required. Variables can be assigned to other variables, reassigned to new values at any moment, and even assigned to another variable. When a new value is assigned to a variable, the previous value is lost and cannot be recovered. This reassignment of a variable is also possible when the data type is not the same. For instance, an integer variable can be reassigned to a text variable and vice versa. This is not true for most of the other programming languages. The last sentence assigns values to two variables in the same statement, which is a feature that is rarely found in any other programming language. Finally, variable names are case sensitive; for example, the variable name "gene_symbol" cannot be recall as "Gene Symbol" or "GENE SYMBOL."

12.4 **Strings**

For computer programmers, strings are the collection of characters or, more commonly, any texts. String manipulation is quite prevalent in bioinformatics investigations, such as sequence files, pattern discovery in sequences, data mining from texts, and data processing from a variety of file types. In Python, a string object can be formed by containing a sequence of characters within a pair of single quotes/double quotes/triple single quotes/triple-double quotes. While characters included in single or double quotation marks can only contain a single line, characters wrapped in triple single or triple-double quotation marks can contain many lines. The characters should be enclosed within the same type of quote—usually single or double quotes for defining a string datatype. Consider the following example (Table 12.3).

12.4.1 String indexing

As a string is a collection or sequence of characters, Python allows for the extraction of individual characters as well as portions of the text via their indexes. To retrieve the character, the index number must be placed following the string variable inside the square bracket pair.

The following is an example of using String indexing to print the second character in the word "PLANT."

It's worth noting that the index of any string begins with 0, beginning with the leftmost character; so, the index of the first character "P" is 0; the index of the second character "L" is 1, and so on. Backward indexing begins with −1 from the rightmost character, indicating that the last character "T" has a backward index of −1, the second last character "N" has a backward index of −2, and so on.

Another example of character index for Python is shown in Table 12.4, where first row is the sequence, the second row is the forward index of nucleotides, and the third row shows the backward index:

Table 12.3 Defining strings in Python.

Code	Comments	Output
seq_1 = 'MALNSGSPPA' print(seq_1)	A string within a pair of single quotes	MALNSGSPPA
seq_2 = "MALNSGSPPA" print(seq_2)	A string within a pair of double quotes	MALNSGSPPA
seq_3 = '''MALNSGSPPA''' print(seq_3)	A string within a pair of triple single quotes	MALNSGSPPA
seq_4 = """MALNSGSPPA""" print(seq_4)	A string within a pair of triple double quotes	MALNSGSPPA
seq_5 = '''MALNSGSPPA IGPYYENHGY''' print(seq_5)	A string within a pair of triple single quotes, can have multiple lines	MALNSGSPPA IGPYYENHGY
seq_6 = """IGPYYENHGY IGPYYENHGY""" print(seq_6)	A string within a pair of triple double quotes, can have multiple lines	IGPYYENHGY IGPYYENHGY

To take a portion of a text or string, use the annotation "string_name [start: end]", where the start is the starting index and the end is the index extending up to but not including it.

*   *word*[1:3] is "LA"—characters starting at index 1 and extending up to but not including index 3
*   *word*[3:] is "NT"—leaving a blank for either index defaults to the start or end index of the string
*   *word*[:] is "PLANT"—emptying both always gives us a copy of the whole thing.
*   *word*[1:10] is "LANT"—an index that is too big is truncated to string length.
*   *word*[:-2] is "PLA"—selecting up to but not including the last 4 chars.
*   *word*[-2:] is "NT"—starting with the fourth character from right end to the right end.

12.4.2 Operations on strings

There are a few ways to concatenate or join strings. The easiest and most common way to add join strings is to use the plus symbol (+). i.e., in simplest terms, merely adding them.

The "+" operator can be used to combine any number of strings. A critical point to remember is that when adding strings, all datatypes must be strings; for example, if users add a string with an integer, such as "PLANT"+4, an error message indicates that the "str" type and the "int" type cannot be added. To add a number, it must first convert it to the "str" type using the str (number) function. While integers and strings cannot be added, the same string can be printed several times using the "*" operator and a "int" datatype. For instance, "PLANT"*2 returns the string twice, i.e., "PLANTPLANT".

12.4.3 Methods in strings

Several string handling methods include count (), find (), and len (). Their application is outlined below in Table 12.5.

In the above instances, "len ()" is a function that returns the string's length. There is a key method called str.split () which is regularly been used to extract data from delimited text file formats such as CSV, TSV, and others. CSV stands for comma-separated values, in which each column's values are separated by a comma, while TSV means for tab-separated values, in which each column's values are separated by a tab delimiter.

Table 12.4 String indexing in Python.

	P	L	A	N	T
Forward indexing	0	1	2	3	4
Backward indexing	−5	−4	−3	−2	−1

Table 12.5 Few methods in strings.

Code	Output
`peptide = 'TSLWGLLFLSAALSLWPTSG'`	
`print(peptide.count('A'))`	2
`print(peptide.find('LW'))`	2
`print(Len(peptide))`	20

```
sepal_length,sepal_width,petal_length,petal_width,species
5.1,3.5,1.4,0.2,Iris-setosa
4.9,3.0,1.4,0.2,Iris-setosa
4.9,2.4,3.3,1.0,Iris-versicolor
5.9,3.0,5.1,1.8,Iris-virginica
```
FIG. 12.1

Example of a CSV formatted file.

Fig. 12.1 is an example of CSV formatted file where the first row is known as header row, which consists of column names, and the rest of the rows are instances having values separated by a comma for each column. The values can be extracted from each row if each row is considered as a string using str.split () method:

In Table 12.6 the first observation, or the second row of the csv file in Fig. 12.3, is assigned to a variable named "first row". Then Python's many variables assignment feature was used to establish each column as a variable and the first observations as its value. The split (",") method returns a list of values separated by commas. The variables named after the CSV file's columns can be printed. The output also includes a list of split values. List is a Python data type that will be covered in the next section. There are additional intriguing methods available for the string data-type, which can be found in the Python documentation.

12.5 Python lists and tuples

After learning about datatypes like integers, strings, and booleans, this section will cover Lists.

Lists, like containers, store many values of any type. They're called data structures because they store data in a way that makes retrieval easy.

Lists are similar to arrays in other programming languages, but they are more versatile. List items are called elements. Lists have crucial features like:

- List keeps track of the items inserted and can be accessed later.
- Index access—a list's objects can be indexed.
- Lists can contain any entity—numbers, strings, and even other lists.
- Lists can be mutated—New items can be added, and old ones can be removed or altered.

Table 12.6 String's split () method.

Code	Output

`#str.split()`
`first_row = '5.1,3.5,1.4,0.2,Iris-setosa'`
`sepal_length,sepal_width,petal_length,petal_widt`
`h,species = first_row.split(',')`
`print(sepal_length)`
`print(sepal_width)`
`print(petal_length)`
`print('---------')`
`print(first_row.split(','))`
	
	`5.1`
	`3.5`
	`1.4`
	`---------`
	`['5.1','3.5','1.4',`
	`'0.2','Iris-setosa']`

12.5.1 Accessing values in list

Like strings, list items also have indexes starting with "0" for forward indexing and "−1" for backward indexing (Fig. 12.2).

The items inside a list can be accessed using brackets [] and indexes.

Slicing a list allows to access a subset of it. The string slice operator can also slice lists. Similarly, to string, omitting the first index causes the slice to begin at the beginning. If the second is absent, the slice ends. If both of them are removed, the slice is a copy of the List (Table 12.7).

The "+" operator can be used to concatenate two lists and the "*" operator to repeat a list any specified number of times.

	['Moss'	'Embryophyte'	'Thallophyte'	'Conifer']
Forward indexing	0	1	2	3
Backward Indexing	-4	-3	-2	-1

FIG. 12.2

Python list indexes.

Table 12.7 List slicing.

Code	Output

# not including index 2 	
print(plants[0:2])
# everything up to index 3 	
print(plants[:3])
# index 1 to end of list 	
print(plants[1:])
# Coping whole list 	
print(plants[:])
	['Moss', 'Embryophyte']
	
	['Moss', 'Embryophyte', 'Thallophyte']
	
	['Embryophyte', 'Thallophyte', 'Conifer']
	
	['Moss', 'Embryophyte', 'Thallophyte', 'Conifer']

12.5.2 Methods with lists

Python provides some in-built methods for List such as:

- count () methods will return the total number of occurrences of an item in the List.
- index () will give the index of an item.
- append () adds an item at the end of the List.
- remove () will remove the first occurrence of the item in the List.
- pop () will remove the item at index provided by the user.
- min (), max () and sum () will provide the minimum, maximum and sum of the lists constituting number values.
- len () will provide the total number of items in the List.
- sort () method can be used to sort a list of numerical values in increasing or decreasing order, or a list of string in A-Z or Z-A order.

12.5.3 Tuples

A "tuple" is one of Python's four built-in data types for storing collections. Tuples' components, unlike those of other data types, are both sequential and immutable. They can be used to keep track of a variety of different items in a single variable, and they come with a number of useful actions pre-programmed in. There are numerous parallels between these tuples and lists. Developers who are familiar with lists, a commonly used data structure, may mistake lists for tuples. Tuples in Python, like lists, are collections of elements of any data type, but unlike lists, tuples are immutable; that is, once assigned, users cannot change the tuple's contents or the tuple itself.

12.6 Dictionary in Python

Dictionaries are data structures in Python that are similar to hash tables or hashmaps in other computer languages. Each key corresponds to a single value in a dictionary. The ideal approach to establish a dictionary is to put the key:value pairs inside curly brackets "{}". Only "{}" can declare an empty dictionary (Fig. 12.3) (Table 12.8).

The Python dictionary has the following properties:

- Dictionaries are unordered—the key-value pairs are not sorted.
- Dictionaries are mutable, which means they can be expanded or trimmed as needed.
- The dictionary values can be access by keys. Indexes can't be accessed because key-value pairs aren't kept in order.
- A dictionary can only contain unique keys. The keys in the dictionary are always unique.
- Dictionary keys, such as strings, integers, and tuples, should be immutable data types.

Python dictionary values can be numbers, texts, lists, or even dictionaries. A nested dictionary is a dictionary inside a dictionary. The key in square bracket retrieves the dictionary's stored values.

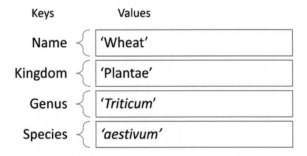

FIG. 12.3

Python dictionary key: Value pairs.

Table 12.8 Creating a Python dictionary.

Code	Output

`crop = {}`
`crop`
`{'Name':'Wheat','Kingdom':'Plantae','Gen us':'Triticum','Species':'aestivum'}`
`print(crop)`
`print(type(crop))`
	`{'Name': 'Wheat', 'Kingdom':`
	`'Plantae', 'Genus':`
	`'Triticum', 'Species':`
	`'aestivum'}`
	`<class 'dict'>`

12.7 **Conditional statements**

Until now, the programmes are simple, not clever, and not making decisions. Conditional statements are required to make a program make decisions based on conditions. Computers have only two states, True or False, like a light switch has two states, On or Off. In Python, these True/False situations are known as booleans.

A condition is always defined by comparison, such as larger than, less than, or equal. Here are some comparisons with Python operators:

- Equal: a == b
- Not Equal: a != b
- Less than: a < b
- Less than or equal to: a <= b
- Greater than: a > b
- Greater than or equal to: a >= b

All these comparisons result in Boolean values "True" or "False".

12.7.1 **Logical operators**

When comparing many conditions, logical operator are used. The logical operators "and", "or", and "not" are the same in Python as in English. Logic operators usually work on conditions.

"and" operator will only give true if both the conditions are true, "or" will give true if either of the conditions is true, lastly "not" will give just the opposite condition, i.e., it will give false for true and true for false.

12.7.2 **If and else statements**

Often, it is required to execute some statement only if some conditions are true. For this, "if" and "else" statements are used.

1. "If" Statement: use it to execute a block of code if the specified condition is true.
2. "Else" statement: use with an "if" statement to execute a block of code if the specified condition is false.

In Python, blocks are defined by indentation after a colon. Indentation is a strict Python syntax for forming a statement block, which can be commonly encounter. If the condition is true, an indented block of statements will be executed (see Fig. 12.4). Let's see an example of comparing the expression of a gene in a controlled and treated environment in Table 12.9.

Next example is for comparing the expression of a gene in a controlled and treated environment, with "If" and "else" statements (Table 12.10).

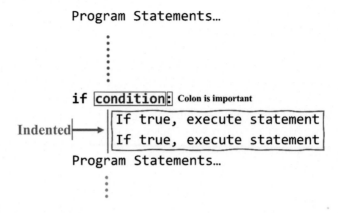

FIG. 12.4

Syntax of "if" statement.

Table 12.9 If statements.

Code	Output

```control_expression = 14```
```treated_expression = 3.5```
```if   control_expression```
```treated_expression:```
```   print('downregulated')```
	downregulated

Table 12.10 If, else statement.

Code	Output

```control_expression = 14```
```treated_expression = 3.5```
```if control_expression > treated_expression:```
```   print('Gene is downregulated')```
```else:```
```   print('Gene is upregulated')```
	Gene is downregulated

12.8 Loops in Python

When developing code, it is occasionally necessary to repeat a block of code or a statement until the condition is met or the statement is false (depend on requirement). This type of recurring or iterative behavior can be implemented using loops. These are known as control structures, and programming languages offer a variety of options to facilitate their implementation.

Generally, there are two type of loop structure:

1. While Loop
2. For Loop

12.8.1 While loop

In Python, a loop is a set of statements that are repeated until a stop condition is met. The while loop syntax includes a stop condition. A while loop works by executing a set of statements until the stop condition is met (Fig. 12.5). A simple example of "while" loop is printing numbers from 0 to 5 in shown in Table 12.11.

In the preceding code (Table 12.14), the variable "a" is initialized at 0. While statements follows with a stop condition of "a" less than 6. While statements comprise two statements, one for printing "a" and the other for incrementing "a" by 1. The value of "a" is printed and incremented from 0 to 1 when the block is first run. The while block's stop condition is checked before rerunning. The block will rerun if "a" = 1 is less than 6. When "a" equals 6, the loop condition fails and the program will stop. In the absence of a stop condition, or a condition that can never be false, the loop becomes infinite and is executed indefinitely. The Jupyter notebook's Python kernel has to be restarted in this scenario. If the asterisk on the left side of the cell persists, them it may be in an infinite loop.

The comments accompany each statement in the preceding example (Table 12.14). The stop condition is the last index in the list in this case. The List's

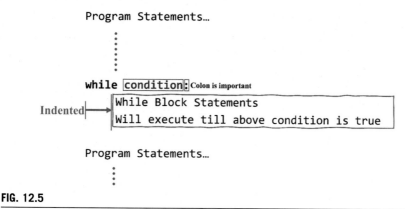

FIG. 12.5

Syntax of a Python while loop.

Table 12.11 Simple while loop.

Code	Output

a = 0
while a<6:
    print(a)
      a = a+1
	0
	1
	2
	3
	4
	5

last index in the forward direction is 5 and its length is 6, as it contains that many members. Thus, the stop condition is satisfied when the final index is increased by one and equals the length of the 'total data' List, i.e. when both are equal to six.

12.8.2 "For" loop

Loops are indefinite loops because they run until a particular condition is false. However, often users need to loop through words in texts or items in lists or through keys and values of a dictionary. In such cases, they require definite loops like "for" loop for doing these tasks of iterating over a sequence datatypes. The "for" loop runs through each item in a set of items. The syntax of a "for" loop is similar as "while" loop: there is a "for" statement and a block of code under that statement (Table 12.12).

The "for" loop iterates over each item in the List and runs the block statements for all items in the List. In simple English, this for loop can be translated as, run the block of code under the "for" loop for every plant in the plants list. In the "for" statement, the "plant" is the variable which changes with every.

12.8.3 Breaking a loop

The "break" command in Python is used to exit a loop early. With a "break" command, the program exits the loop immediately and executes the remaining statements outside the loop block (Table 12.13).

Table 12.12 Simple "for" loop.

Code	Output
plants = ['Moss', 'Embryophyte', 'Thallophyte','Conifer'] for plant in plants: print(plant)	Moss Embryophyte Thallophyte Conifer

Table 12.13 Python code snippets showing the breaking a loop before its ending.

Code	Output
```for temp in plants: print(temp) if temp == 'Thallophyte': break```	```Moss Embryophyte Thallophyte```

As the "if" condition is satisfied when "temp" variable equals to "Thallophyte", the execution of "break" statement blocks the printing of "Conifer".

12.9 File handling in Python

Until now, our computer's primary memory has been accessible while constructing and executing statements and programmes. Our computer, as previously stated, has primary and secondary memory, with primary memory being erased when the system is turned off. Secondary memory, such as a hard drive, can save data even after the system has been turned off. Files are often stored on hard drives and have unique paths allocated to them. Biological data is also saved in files with specialized formats such as PDB, networks, and sequence files. As a result, being able to access and manipulate files is crucial for a research project. Files are also used to store and share enormous amounts of data, which is why learning how to read a file in Python is vital.

The first step in file handling is to use Python's built-in "open ()" function to open a file. Then instruct the operating system to identify and verify the presence of the file by calling the open () function.

Code:

```
f = open('myfile.txt') # If file is in current directory
f = open('C:\Python33\Scripts\myfile.txt') # opening file with Exact Path
```

When only the filename is specified, it is assumed that the file is in the same folder as Python. As illustrated in the second line of code, we may also indicate the file's location by its full path.

12.9.1 **Specify file mode**

There are eight types of modes for file operation (Table 12.14).

The FASTA file format is used to contain both gene IDs and sequences. After the identifier, the ID line/first line can have further information. It always starts with a ">." The order of the genes is written after the first line. In the example below (Table 12.15), The targeted file named "myfile.fasta" was opened and read which is located in the current directory or folder. The following is the file's content:

>ID001 (gene abcb1)

GATATGATGCCGTCACTA

GTTTACCGCTGGTAACTG

Here "open ()" is a class which deals with operations of files (Table 12.15). The "open" mode is initiated with the name or location of the file. This class has a method called read () which reads all contents of the file. Open handler has a method for returning the List of all the lines known as readlines (). Users can write within a file by opening it in write mode and utilizing the handler class open () and write () function (Table 12.16).

The new file will be saved and can be found in the current folder or directory. The contents of the file will be:

```
>ID002 (gene TLA)
ATGCTTTGGCCAAATTGG
GGTTCCATGGTCATGC
TGCTGATC
```

Table 12.14 Modes of file operation in Python.

Character	Mode	Description
"r"	Read (default)	Open a file for read only
"w"	Write	Open a file for write only (overwrite)
"a"	Append	Open a file for write only (append)
"r+"	Read + Write	Open a file for both reading and writing
"x"	Create	Create a new file
"t"	Text (default)	Read and write strings from and to the file.
"b"	Binary	Read and write bytes objects from and to the file. This mode is used for all files that don't contain text (e.g., images).

Table 12.15 Reading from a file using Python.

Code	Output
f = open('my_file.fasta') print(f.read())	>ID001 (gene abcb1) GATATGATGCCGTCACTA GTTTACCGCTGGTAACTG

Table 12.16 Writing inside a file.

Code	Output
f = open('my_new_file.fasta', 'w') f.write('>ID002 (gene TLA)\n') lines = ['ATGCTTTGGCCAAATTGG\n', 'GGTTCCATGGTCATGC\n', 'TGCTGATC'] f.writelines(lines)	

When a file is opened, it must be closed with the close () methods of the "f" object. f.close () will close a file.

12.10 Importing functions

A script with the.py extension is a Python module. Python modules can be written relatively quickly. Create a file containing valid Python code, then give it a name and the.py extension. Python's modules make it easier to modularize programming. When creating Python applications, modules may be accessed by using the "import" keyword. Program will be more dependable and effective by using modules.

For practically every programming task, including web development, database construction, image analysis, data science, statistics, machine learning, and more, Python packages and modules are available. Libraries are a collection of functions, whereas packages are a collection of modules. Over 227,607 Python packages are presently available in the Python Package Index (PyPI), which was created to make life simpler for developers. Using the "pip" installer, we may install any of the PyPI packages that are offered. Both the "conda" and "pip" installers are included in the Python anaconda distribution. Open the terminal or command prompt and run "pip install PackageName" or, for the Anaconda distribution, "conda install PackageName" to install a package. Numerous preloaded packages for data science applications are included in the Python Anaconda distribution. Using the keywords "from" and "import," we may call the modules once the packages have been installed.

12.10.1 Running a t-test in Python

The relationship between the means of two groups may be deduced using an inferential statistic known as a t-test. T-tests are used when data sets have a normal distribution but unknown variances, such as the a set produced by flipping a coin 100 times.

Assume a pharmaceutical company is testing a brand-new medicine. One group of patients is given the treatment, while another, the "control group," is given a placebo. A group is given a placebo, which is a substance that doesn't work as a drug, to see how another group reacts to the real drug. After the drug trial, those in the control group who were given a placebo said their average life expectancy went up by 3 years, while those in the group that got the new therapy said it went up by 4 years. Initial trials indicate that the drug is effective. The observation may, however, have been made by chance. T-test can be used to determine if the findings are consistent across the board.

Assume that we are conducting an experiment in which we expect to accelerate plant development by changing the concentration of a certain nutrient in one plot (the "treatment plot") while maintaining the same concentration in another and comparing the results between them. To test the difference in plant height between control and treatment plots, we can measure the height randomly from a few samples in each plot. Even if our nutrient has no effect on the plants in the treatment plot, we can still identify random variances in our data (plants never grow at precisely the same rate). If the differences between the plots are too large to be explained by chance, we may use the t-test to find out (Table 12.17).

This yields two numbers: a T statistic indicating how different our plant sizes are between treatments and a *P*-value indicating the likelihood of obtaining such a large difference by chance if our treatment was ineffective.

Even though the plants in our treatment plot grew larger on average, this t-test result indicates that the difference could easily be due to chance. Because *P* is around 0.03, there is a 3% chance that this result is a coincidence. Differences with a less than 5% chance of occurring by chance are considered statistically significant. So our difference in this example is statistically significant.

12.10.2 Make a simple scatterplot in matplotlib

To attract mates, male frogs frequently make loud calls. Assume we were studying frog calls and wanted to see if larger frogs made longer calls than smaller frogs. Assume we measured the length and size of several frogs' calls. We'd like to create a

Table 12.17 T-test in Python using scipy library.

Code	Output

#Finding the stats module in scipy and importing #its ttest function.
from scipy.stats import ttest_ind
 	
# control and treatment plot in centimetres. 	
control_plant_sizes = [4,7,8,3,4,2,2]
 	
treatment_plant_sizes = [5,4,10,8,5,12,13]
 	
#Run a t-test on the results
ttest_ind(control_plant_sizes,treatment_plant_s izes)
	Ttest_indResult(stat istic=- 2.35900095298480, pvalue=0.03611687624 692063)

graph that depicts the relationship between frog size and call length. A scatterplot, on the other hand, could be a good way to visualize this data.

A scatterplot of frog sizes and frog call durations may be generated by importing the appropriate functions, which include labeling the scatterplot, x and y axes form matplotlib library of python. Matplotlib is a low level and popular graph plotting library in python (Table 12.18 and Fig. 12.6).

Table 12.18 Plotting scatter plot in Python using matplotlib library.

Code

from matplotlib.pyplot import scatter, xlabel, ylabel
 	
# frog sizes in grams 	
frog_sizes = [155,475,260,525,300,280,315]
 	
# call lengths of those same frongs, in seconds 	
frog_call_lengths = [1,5,2,6,3,3,4]
scatter(frog_sizes,frog_call_lengths)
xlabel("Frog Mass")
ylabel("Call Length")

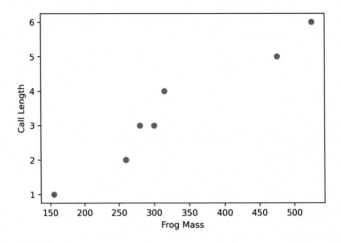

FIG. 12.6

Scatterplot plotted using matplotlib library of Python, where x denotes is frog mass and y axis denotes call lengths.

12.10.3 Running a simple linear regression in Python

The t-test function in the previous section came from the scipy.stats module, which includes a wide range of statistical analyses. The pearson r function, which performs Pearson regression, is an example of this. This compares the values of two sets of numbers to see if there is any relationship between them. Thus, while t-tests can be used to compare values within a category, Pearson regression can be used to compare two continuous variables.

We can use a Pearson regression to see if larger frogs have longer calls now that we know how long their calls are (Table 12.19).

12.11 Data handling

Pandas is a high-level data manipulation and analysis package. It is written in Python and is available for free to anyone. Python's pandas library makes it simple to load, format, edit, and inspect data. It is a popular option among developers and is utilized for a variety of data science projects. Pandas are fantastic since they are simple to use and operate quickly. Users can process up to 10 million rows of data without issue. A pandas Dataframe is simple to modify since one can simply add or delete columns, slice it, index it, or deal with missing data. Using pandas makes it easy to clean, modify, and analyze data. One can delve into a dataset stored on their computer as a comma-separated values (CSV) file. Pandas will read the CSV file and convert information to a DataFrame, which is similar to a tables in excel from which the user can perform actions such as:

Table 12.19 Regression analysis using Python.

Code	Output
``#Import the pearsonr function
from scipy.stats import pearsonr
#Run pearson regression
R2,   pvalue =
pearsonr(frog_sizes,frog_call_lengths)

#Print the results
print("R2:",R2)
print("pvalue:",pvalue)``` | R2:
0.9704929909196073

pvalue:
0.0002827268106013687 |

- Answer queries like "What is the average, median, highest, and lowest value for each column?" via statistical analysis.
- Is there anything that connects rows A and B?
- How can data in a column frequently come apart?
- Remove any incorrect information and sort the data by selecting which rows or columns to examine.
- Use Matplotlib to transform the data into something visual. Anything can be drawn, from lines and bars to bubbles and histograms.
- Return the modified data to a CSV file, a database, or another storage location.

What has been covered in this chapter is only the tip of the iceberg in terms of what Python can do. In any event, the ideas presented here may inspire everyone to experiment with one of their own data sets. It's simple to go to libraries like Matplotlib, seaborn, or scipy from here, which may be used to visualize data, conduct statistical tests on it, or fit data to mathematical models.

The advancements in the Python language are too numerous to describe in this brief introduction. Users will need to learn more about algorithms, mathematical modeling, data visualization, and other topics. Instead of merely undertaking random programming challenges, if readers want to develop anything important with the language, it may provide them more drive to stay focused and find all of the necessary ingredients. Learning to program computers is an important ability that can help us find better employment and deal with modern time's issues. Those interested in learning Python will find a wealth of high-quality resources. Some of these have already been mentioned, and Table 12.20 shows a list of available resources.

Table 12.20 Resources for learning Python.

Source type	Address	Description
Python	https://www.python.org/	Official python website
	http://hplgit.github.io/bioinf-py/doc/pub/html/index.html	Illustrating python via examples from bioinformatics
Codes of published research articles	https://paperswithcode.com/	Open resource with machine learning papers, code, datasets, methods and evaluation tables.
Biopython	https://biopython.org/; http://biopython.org/DIST/docs/tutorial/Tutorial.html	The Biopython project is an international collaboration of scientists working to create open-source python programmes for use in computational molecular biology.
Visualization of Genomic data	https://gangcaolab.github.io/CoolBox/index.html	Users can use CoolBox to create high-quality visualization plots and analyze their data in a flexible, customizable, and user-friendly manner. Molecular computational biology

References

Ekmekci, B., McAnany, C.E., Mura, C., 2016. An introduction to programming for bioscientists: A python-based primer. PLoS Comput. Biol. 12, e1004867. https://doi.org/10.1371/JOURNAL.PCBI.1004867.

Hasija, Y., Chakraborty, R., 2021. Hands on Data Science for Biologists Using Python. CRC Press.

Marx, V., 2013. The big challenges of big data. Nature 7453 (498), 255–260. https://doi.org/10.1038/498255a.

Python 3.0 Release | Python.org [WWW Document], (n.d.). URL https://www.python.org/download/releases/3.0/ (accessed 8.21.2022).

van Rossum, G., Drake, F.L., 2009. Python 3 Reference Manual. CreateSpace, Scotts Valley, CA.

Index

Note: 'Page numbers followed by "*f*" indicate figures and "*t*" indicate tables.'